Now into its sixth edition and with an additional chapter on computers and number theory, written by J. H. Davenport, *The Higher Arithmetic* introduces concepts and theorems in a way that does not require the reader to have an in depth knowledge of the theory of numbers but also touches upon matters of deep mathematical significance.

'Although this book is not written as a textbook but rather as a work for the general reader, it could certainly be used as a textbook for an undergraduate course in number theory and, in the reviewers opinion, is far superior for this purpose to any other book in English.'
From a review of the first edition in *Bulletin of the American Mathematical Society.*

THE HIGHER
ARITHMETIC

AN INTRODUCTION TO
THE THEORY OF NUMBERS

Sixth edition

H. Davenport

M.A., SC.D., F.R.S.

late Rouse Ball Professor of Mathematics
in the University of Cambridge and
Fellow of Trinity College

CAMBRIDGE
UNIVERSITY PRESS

Published by the Press Syndicate of the University of Cambridge
The Pitt Building, Trumpington Street, Cambridge CB2 1RP
40 West 20th Street, New York, NY 10011-4211, USA
10 Stamford Road, Oakleigh, Victoria 3166, Australia

First published by Hutchinson & Co Ltd 1952
Fifth edition published by Cambridge University Press 1982
Sixth edition published by Cambridge University Press 1992

Printed in Great Britain at The Bath Press, Avon

A catalogue record for this book is available from the British Library

Library of Congress cataloguing in publication data available

ISBN 0 521 41998 0   hardback
ISBN 0 521 42227 2   paperback

CONTENTS

INTRODUCTION

The higher arithmetic, or the theory of numbers, is concerned with the properties of the natural numbers 1, 2, 3, ... These numbers must have exercised human curiosity from a very early period; and in all the records of ancient civilizations there is evidence of some preoccupation with arithmetic over and above the needs of everyday life. But as a systematic and independent science, the higher arithmetic is entirely a creation of modern times, and can be said to date from the discoveries of Fermat (1601–1665).

A peculiarity of the higher arithmetic is the great difficulty which has often been experienced in proving simple general theorems which had been suggested quite naturally by numerical evidence. 'It is just this,' said Gauss, 'which gives the higher arithmetic that magical charm which has made it the favourite science of the greatest mathematicians, not to mention its inexhaustible wealth, wherein it so greatly surpasses other parts of mathematics.'

The theory of numbers is generally considered to be the 'purest' branch of pure mathematics. It certainly has very few direct applications to other sciences, but it has one feature in common with them, namely the inspiration which it derives from *experiment*, which takes the form of testing possible general theorems by numerical examples. Such experiment, though necessary in some form to progress in every part of mathematics, has played a greater part in the development of the theory of numbers than elsewhere; for in other branches of mathematics the evidence found in this way is too often fragmentary and misleading.

As regards the present book, the author is well aware that it will not be read without effort by those who are not, in some sense at least, mathematicians. But the difficulty is partly that of the subject itself. It cannot be evaded by using imperfect analogies, or by presenting the proofs in a way which may convey the main idea of the argument, but is inaccurate in detail. The theory of numbers is by its nature the most exact of all the sciences, and demands exactness of thought and exposition from its devotees.

The theorems and their proofs are often illustrated by numerical examples. These are generally of a very simple kind, and may be despised by those who enjoy numerical calculation. But the function of these examples is solely to illustrate the general theory, and the question of how arithmetical calculations can most effectively be carried out is beyond the scope of this book.

The author is indebted to many friends, and most of all to Professor Erdös, Professor Mordell and Professor Rogers, for suggestions and corrections. He is also indebted to Captain Draim for permission to include an account of his algorithm.

The material for the fifth edition was prepared by Professor D. J. Lewis and Dr J. H. Davenport. The problems and answers are based on the suggestions of Professor R. K. Guy.

Chapter VIII and the associated exercises were written for the sixth edition by Professor J. H. Davenport.

I

FACTORIZATION AND THE PRIMES

1. *The laws of arithmetic*

The object of the higher arithmetic is to discover and to establish general propositions concerning the natural numbers 1, 2, 3, ... of ordinary arithmetic. Examples of such propositions are the fundamental theorem (I.4)* that *every natural number can be factorized into prime numbers in one and only one way*, and Lagrange's theorem (V.4) that *every natural number can be expressed as a sum of four or fewer perfect squares*. We are not concerned with numerical calculations, except as illustrative examples, nor are we much concerned with numerical curiosities except where they are relevant to general propositions.

We learn arithmetic experimentally in early childhood by playing with objects such as beads or marbles. We first learn addition by combining two sets of objects into a single set, and later we learn multiplication, in the form of repeated addition. Gradually we learn how to calculate with numbers, and we become familiar with the laws of arithmetic: laws which probably carry more conviction to our minds than any other propositions in the whole range of human knowledge.

The higher arithmetic is a deductive science, based on the laws of arithmetic which we all know, though we may never have seen them formulated in general terms. They can be expressed as follows.

Addition. Any two natural numbers a and b have a *sum*, denoted by $a+b$, which is itself a natural number. The operation of addition satisfies the two laws:

*References in this form are to chapters and sections of chapters of this book.

$$a+b=b+a \qquad (commutative\ law\ of\ addition),$$
$$a+(b+c)=(a+b)+c \qquad (associative\ law\ of\ addition),$$

the brackets in the last formula serving to indicate the way in which the operations are carried out.

Multiplication. Any two natural numbers a and b have a *product*, denoted by $a \times b$ or ab, which is itself a natural number. The operation of multiplication satisfies the two laws

$$ab=ba \qquad (commutative\ law\ of\ multiplication),$$
$$a(bc)=(ab)c \qquad (associative\ law\ of\ multiplication).$$

There is also a law which involves operations both of addition and of multiplication:

$$a(b+c)=ab+ac \qquad (the\ distributive\ law).$$

Order. If a and b are any two natural numbers, then either a is equal to b or a is *less than* b or b is *less than* a, and of these three possibilities exactly one must occur. The statement that a is less than b is expressed symbolically by $a<b$, and when this is the case we also say that b is greater than a, expressed by $b>a$. The fundamental law governing this notion of order is that

$$if \quad a<b \quad and \quad b<c \quad then \quad a<c.$$

There are also two other laws which connect the notion of order with the operations of addition and multiplication. They are that

$$if \quad a<b \quad then \quad a+c<b+c \quad and \quad ac<bc$$

for any natural number c.

Cancellation. There are two laws of cancellation which, though they follow logically from the laws of order which have just been stated, are important enough to be formulated explicitly. The first is that

$$if \quad a+x=a+y \quad then \quad x=y.$$

This follows from the fact that if $x<y$ then $a+x<a+y$, which is contrary to the hypothesis, and similarly it is impossible that $y<x$, and therefore $x=y$. In the same way we get the second law of cancellation, which states that

$$if \quad ax=ay \quad then \quad x=y.$$

Subtraction. To subtract a number b from a number a means to find, if possible, a number x such that $b+x=a$. The possibility of subtraction is related to the notion of order by the law that *b can be subtracted from a if and only if b is less than a.* It follows from the first cancellation law that if subtraction is possible, the resulting number is unique; for if $b+x=a$ and $b+y=a$ we get $x=y$. The result of subtracting *b* from *a* is denoted by $a-b$. Rules for operating with the minus sign, such as $a-(b-c)=a-b+c$, follow from the definition of subtraction and the commutative and associative laws of addition.

Division. To divide a number *a* by a number *b* means to find, if possible, a number *x* such that $bx=a$. If such a number exists it is denoted by $\dfrac{a}{b}$ or a/b. It follows from the second cancellation law that if division is possible the resulting number is unique.

All the laws set out above become more or less obvious when one gives addition and multiplication their primitive meanings as operations on sets of objects. For example, the commutative law of multiplication becomes obvious when one thinks of objects arranged in a rectangular pattern with *a* rows and *b* columns (fig. 1); the total number of objects is *ab* and is also *ba*. The distributive law becomes obvious when one considers the arrangement of objects indicated in fig. 2; there are $a(b+c)$ objects

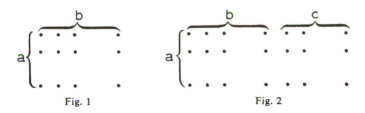

Fig. 1   Fig. 2

altogether and these are made up of *ab* objects together with *ac* more objects. Rather less obvious, perhaps, is the associative law of multiplication, which asserts that $a(bc)=(ab)c$. To make this apparent, consider the same rectangle as in fig. 1, but replace each object by the number *c*. Then the sum of all the numbers in any one row is *bc*, and as there are *a* rows the total sum is $a(bc)$. On the other hand, there are altogether *ab* numbers each of which is *c*, and therefore the total sum is $(ab)c$. It follows that $a(bc)=(ab)c$, as stated.

The laws of arithmetic, supplemented by the principle of induction (which we shall discuss in the next section), form the basis for the logical development of the theory of numbers. They allow us to prove general theorems about the natural numbers without it being necessary to go back to the primitive meanings of the numbers and of the operations carried out on them. Some quite advanced results in the theory of numbers, it is true, are most easily proved by counting the same collection of things in two different ways, but there are not very many such.

Although the laws of arithmetic form the logical basis for the theory of numbers (as indeed they do for most of mathematics), it would be extremely tedious to refer back to them for each step of every argument, and we shall in fact assume that the reader already has some knowledge of elementary mathematics. We have set out the laws in detail in order to show where the subject really begins.

We conclude this section by discussing briefly the relationship between the system of natural numbers and two other number-systems that are important in the higher arithmetic and in mathematics generally, namely the *system of all integers* and the *system of all rational numbers*.

The operations of addition and multiplication can always be carried out, but those of subtraction and division cannot always be carried out within the natural number system. It is to overcome the limited possibility of subtraction that there have been introduced into mathematics the number 0 and the negative integers $-1, -2, \ldots$. These, together with the natural numbers, form the system of all integers:

$$\ldots, -2, -1, 0, 1, 2, \ldots,$$

within which subtraction is always possible, with a unique result. One learns in elementary algebra how to define multiplication in this extended number-system, by the 'rule of signs', in such a way that the laws of arithmetic governing addition and multiplication remain valid. The notion of order also extends in such a way that the laws governing it remain valid, with one exception: the law that if $a < b$ then $ac < bc$ remains true only if c is positive. This involves an alteration in the second cancellation law, which is only true in the extended system if the factor cancelled is not 0:

$$\text{if} \quad ax = ay \quad \text{then} \quad x = y, \quad \text{provided that} \quad a \neq 0.$$

Thus the integers (positive, negative and zero) satisfy the same laws of arithmetic as the natural numbers except that subtraction is now always possible, and that the law of order and the second cancellation law are modified as just stated. The natural numbers can now be described as the *positive integers*.

Let us return to the natural numbers. As we all know, it is not always possible to divide one natural number by another, with a result which is itself a natural number. If it is possible to divide a natural number b by a natural number a within the system, we say that a is a *factor* or *divisor* of b, or that b is a *multiple* of a. All these express the same thing. As illustrations of the definition, we note that 1 is a factor of every number, and that a is itself a factor of a (the quotient being 1). As another illustration, we observe that the numbers divisible by 2 are the even numbers 2, 4, 6, ... , and those not divisible by 2 are the odd numbers 1, 3, 5, ...

The notion of divisibility is one that is peculiar to the theory of numbers, and to a few other branches of mathematics that are closely related to the theory of numbers. In this first chapter we shall consider various questions concerning divisibility which arise directly out of the definition. For the moment, we merely note a few obvious facts.

(i) *If a divides b then $a \leqq b$* (that is, a is either less than or equal to b). For $b = ax$, so that $b - a = a(x - 1)$, and here $x - 1$ is either 0 or a natural number.

(ii) *If a divides b and b divides c then a divides c.* For $b = ax$ and $c = by$, whence $c = a(xy)$, where x and y denote natural numbers.

(iii) *If two numbers b and c are both divisible by a, then $b + c$ and $b - c$ (if $c < b$) are also divisible by a.* For $b = ax$ and $c = ay$, whence

$$b + c = a(x + y) \quad \text{and} \quad b - c = a(x - y).$$

There is no need to impose the restriction that $b > c$ when considering $b - c$ in the last proposition, if we extend the notion of divisibility to the integers as a whole in the obvious way: an integer b is said to be divisible by a natural number a if the quotient $\dfrac{b}{a}$ is an integer. Thus a negative integer $-b$ is divisible by a if and only if b is divisible by a. Note that 0 is divisible by every natural number, since the quotient is the integer 0.

(iv) *If two integers b and c are both divisible by the natural number a, then every integer that is expressible in the form $ub + vc$, where u and v are integers, is also divisible by a.* For $b = ax$ and $c = ay$, whence $ub + vc = (ux + vy)a$. This result includes those stated in (iii) as special cases; if we take u and v to be 1 we get $b+c$, and if we take u to be 1 and v to be -1 we get $b-c$.

Just as the limitation on the possibility of subtraction can be removed by enlarging the natural number system through the introduction of 0 and the negative integers, so also the limitation on the possibility of division can be removed by enlarging the natural number system through the introduction of all positive fractions, that is, all fractions $\dfrac{a}{b}$, where a and b are natural numbers. If both methods of extension are combined, we get the *system of rational numbers*, comprising all integers and all fractions, both positive and negative. In this system of numbers, all four operations of arithmetic—addition, multiplication, subtraction and division—can be carried out without limitation, except that division by zero is necessarily excluded.

The main concern of the theory of numbers is with the natural numbers. But it is often convenient to work in the system of all integers or in the system of rational numbers. It is, of course, important that the reader, when following any particular train of reasoning, should note carefully what kinds of numbers are represented by the various symbols.

2. Proof by induction

Most of the propositions of the theory of numbers make some assertion about every natural number; for example Lagrange's theorem asserts that every natural number is representable as the sum of at most four squares. How can we prove that an assertion is true for *every natural number*? There are, of course, some assertions that follow directly from the laws of arithmetic, as for instance algebraic identities like

$$(n+1)^2 = n^2 + 2n + 1.$$

But the more interesting and more genuinely arithmetical propositions are not of this simple kind.

It is plain that we can never prove a general proposition by verifying that it is true when the number in question is 1 or 2 or 3, and so on, because we cannot carry out infinitely many verifications. Even if we verify that a proposition is true for every number up to a million, or a million million, we are no nearer to establishing that it is true always. In fact it has sometimes happened that propositions in the theory of numbers, suggested by extensive numerical evidence, have proved to be wide of the truth.

It may be, however, that we can find a *general argument* by which we can prove that *if* the proposition in question is true for all the numbers

$$1, 2, 3, \ldots, n-1,$$

then it is true for the next number, n. If we have such an argument, then the fact that the proposition is true for the number 1 will imply that it is true for the next number, 2; and then the fact that it is true for the numbers 1 and 2 will imply that it is true for the number 3, and so on indefinitely. The proposition will therefore be true for every natural number if it is true for the number 1.

This is the principle of proof by induction. The principle relates to propositions which assert that something is true for every natural number, and in order to apply the principle we need to prove two things: first, that the assertion in question is true for the number 1, and secondly that *if* the assertion is true for each of the numbers $1, 2, 3, \ldots, n-1$ preceding any number n, *then* it is true for the number n. Under these circumstances we conclude that the proposition is true for every natural number.

A simple example will illustrate the principle. Suppose we examine the sum $1+3+5+\ldots$ of the successive odd numbers, up to any particular one. We may notice that

$$1=1^2, \quad 1+3=2^2, \quad 1+3+5=3^2, \quad 1+3+5+7=4^2,$$

and so on. This suggests the general proposition that *for every natural number n, the sum of the first n odd numbers is n^2*. Let us prove this general proposition by induction. It is certainly true when n is 1. Now we have to prove that the result is true for any number n, and by the principle of induction we are entitled to suppose that it is already known to be true for any number less than n. In particular, therefore, we are entitled to suppose that we already know that the sum of the first $n-1$ odd numbers is $(n-1)^2$. The sum of the first n odd numbers is obtained from this by adding the nth odd number, which is $2n-1$. So the sum of the first n odd numbers is

$$(n-1)^2+(2n-1),$$

which is in fact n^2. This proves the proposition generally.

Proofs by induction are sometimes puzzling to the inexperienced, who are liable to complain that 'you are assuming the proposition that is to be proved'. The fact is, of course, that a proposition of the kind now under consideration is a proposition with an infinity of cases, one for each of the natural numbers 1, 2, 3, ...; and all that the principle of induction allows us to do is to suppose, when proving any one case, that the preceding cases have already been settled.

Some care is called for in expressing a proof by induction in a form which will not cause confusion. In the example above, the proposition in question was that *the sum of the first n odd numbers is n^2*. Here n is any one of the natural numbers, and, of course, the statement means just the same if we change n into any other symbol, provided we use the same symbol in the two places where it occurs. But once we have embarked on the proof, n becomes a particular number, and we are then in danger of using the same symbol in two senses, and even of writing such nonsense as 'the proposition is true when n is $n-1$'. The proper course is to use different symbols where necessary.

From a commonsense point of view, nothing can be more obvious than the validity of proof by induction. Nevertheless it is possible to debate whether the principle is in the nature of a *definition* or a *postulate* or an *act of faith*. What seems at any rate plain is that the principle of induction is essentially a statement of the rule by which we enumerate the natural numbers in order: having enumerated the numbers 1, 2, ... , $n-1$ we continue the enumeration with the next number n. Thus the principle is in effect an explanation of what is meant by the words 'and so on', which must occur whenever we attempt to enumerate the natural numbers.

3. *Prime numbers*

Obviously any natural number a is divisible by 1 (the quotient being a) and by a (the quotient being 1). A factor of a other than 1 or a is called a *proper* factor. We all know that there are some numbers which have *no* proper factors, and these are called prime numbers, or *primes*. The first few primes are

$$2, 3, 5, 7, 11, 13, 17, 19, 23, 29, 31, \dots .$$

Whether 1 should be counted as a prime or not is a matter of convention, but it is simpler (as we shall see later) not to count 1 as a prime.

A number which is neither 1 nor a prime is said to be *composite*; such a number is representable as the product of two numbers, each greater than 1. It is well known that by continued factorization one can eventually express any composite number as a product of primes, some of which may of course be repeated. For example, if we take the number 666, this has the obvious factor 2, and we get $666 = 2 \times 333$. Now 333 has the obvious factor 3, and $333 = 3 \times 111$. Again 111 has the factor 3, and $111 = 3 \times 37$. Hence

$$666 = 2 \times 3 \times 3 \times 37,$$

and this is a representation of the composite number 666 as a product of primes. The general proposition is that any composite number is representable as a product of primes. Or, what comes to the same thing, *any number greater than 1 is either a prime or is expressible as a product of primes.*

To prove this general proposition, we use the method of induction. In proving the statement for a number n, we are entitled to assume that it has already been proved for any number less than n. If n is a prime, there is nothing to prove. If n is composite, it can be represented as ab, where a and b are both greater than 1 and less than n. We know that a and b are either primes or are expressible as products of primes, and on substituting for them we get n expressed as a product of primes. This proof is indeed so simple that the reader may think it quite superfluous. But the next general proposition on factorization into primes will not be so easily proved.

The series 2, 3, 5, 7, ... of primes has always exercised human curiosity, and later we shall mention some of the results that are known about it. For the moment, we content ourselves with proving, following Euclid (Book IX, Prop. 20), that *the series of primes never comes to an end.* His proof is a model of simplicity and elegance. Let 2, 3, 5, ... , P be the series of primes up to a particular prime P. Consider the number obtained by multiplying all these primes together, and then adding 1, that is

$$N = 2 \times 3 \times 5 \times \ ... \ \times P + 1.$$

This number cannot be divisible by 2, for then both the numbers N and $2 \times 3 \times 5 \times \ ... \ \times P$ would be divisible by 2, and therefore their difference would be divisible by 2. This difference is 1, and is not divisible by 2. In the same way, we see that N cannot be divisible by 3 or by 5 or by any of the primes up to and including

P. On the other hand, *N* is divisible by *some* prime (namely *N* itself if *N* is a prime, or any prime factor of *N* if *N* is composite). Hence there exists a prime which is different from any of the primes 2, 3, 5, ... , *P*, and so is greater than *P*. Consequently the series of primes never comes to an end.

4. The fundamental theorem of arithmetic

It was proved in the preceding section that any composite number is expressible as a product of primes. As an illustration, we factorized 666 and obtained

$$666 = 2 \times 3 \times 3 \times 37.$$

A question of fundamental importance now suggests itself. Is such a factorization into primes possible in more than one way? (It is to be understood, of course, that two representations which differ merely in the order of the factors are to be considered as the same, e.g. the representation $3 \times 2 \times 37 \times 3$ is to be considered the same as that printed above.) Can we conceive that 666, for example, has some other representation as a product of primes? The reader who has no knowledge of the theory of numbers will probably have a strong feeling that no other representation is possible, but he will not find it a very easy matter to construct a satisfactory general proof.

It is convenient to express the proposition in a form in which it applies to all natural numbers, and not only to composite numbers. If a number is itself a prime, we make the convention that it is to be regarded as a 'product' of primes, where the 'product' has only one factor, namely the number itself. We can go even a stage further, and regard the number 1 as an 'empty' product of primes, making the convention that the value of an empty product is deemed to be 1. This is a convention which is useful not only here but throughout mathematics, since it permits the inclusion in general theorems of special cases which would otherwise have to be excluded, or provided for by a more complicated enunciation.

With these conventions, the general proposition is that *any natural number can be represented in one and only one way as a product of primes*. This is the so-called *fundamental theorem of arithmetic*, and its history is strangely obscure. It does not figure in Euclid's *Elements*, though some of the arithmetical propositions in Book VII of the *Elements* are almost equivalent to it. Nor is it stated explicitly even in Legendre's *Essai sur la théorie des nombres* of 1798. The first clear statement and proof seem to have been given by Gauss in his famous *Disquisitiones Arithmeticae* of 1801. Perhaps the omission of the theorem from Euclid explains why it is passed over without explanation in many schoolbooks. One of them (still in use) describes it as a 'law of thought', which it certainly is not.

We now give a direct proof of the uniqueness of factorization into primes. Later (in §7) we shall give another proof, which will be entirely independent of the present one.

First there is a preliminary remark to be made. *If* the factorization of a particular number m into primes is unique, each prime factor of m must occur in that factorization. For if p is any prime which divides m, we have $m=pm'$, where m' is some other number, and if we now factorize m' into primes we obtain a factorization of m into primes by simply putting on the additional factor p. Since there is supposed to be only one factorization of m into primes, p must occur in it.

We prove the uniqueness of factorization by induction. This requires us to prove it for any number n, on the assumption that it is already established for all numbers less than n. If n is itself a prime, there is nothing to prove. Suppose, then, that n is composite, and has two different representations as products of primes, say

$$n=pqr \dots =p'q'r' \dots ,$$

where p, q, r, ... and p', q', r', ... are all primes. The same prime cannot occur in both representations, for if it did we could cancel it and get two different representations of a smaller number, which is contrary to the inductive hypothesis.

We can suppose without loss of generality that p is the least of the primes occurring in the first factorization. Since n is composite, there is at least one prime besides p in the factorization, and therefore $n \geq p^2$. Similarly $n \geq p'^2$. Since p and p' are not the same, one at least of these two inequalities must be true with strict inequality, and it follows that $pp' < n$. Now consider the number $n-pp'$. This is a natural number less than n, and so can be expressed as a product of primes in one and only one way. Since p divides n it also divides $n-pp'$, and therefore by the preliminary remark it must occur in the factorization of $n-pp'$. Similarly p' must occur. Hence the factorization of $n-pp'$ into primes must take the form

$$n-pp'=pp'\ QR \ldots ,$$

where Q, R, ... are primes. This implies that the number pp' is a factor of n. But $n=pqr$ ... , so it follows on cancelling p that p' is a factor of qr ... . This is impossible by the preliminary remark, because qr ... is a number less than n, and p' is not one of the primes q, r, ... occurring in its factorization. This contradiction proves that n has only one factorization into primes.

The reader will probably agree that the proof, although not very long or difficult, has a certain subtlety. The same is true of other direct proofs of the uniqueness of factorization, of which there are several, all based on much the same ideas. It is important to observe that while the *possibility* of factorization into primes follows at once from the definition of a prime, the proof that the factorization is *unique* is not so immediate. The following illustration, given by Hilbert, explains why these two propositions are on such a different footing from one another.

The definitions of factors and primes involve solely the operation of multiplication, and have no reference to that of addition. Now consider what happens when the same definitions are applied to a system of numbers which can be multiplied together, but which cannot be added or subtracted without going outside the system. Take the system of numbers

$$1, 5, 9, 13, 17, 21, 25, 29, \ldots ,$$

comprising all numbers of the form $4x+1$. The product of any two such numbers is again a number of the same kind. Let us define a 'pseudo-prime' to be a number in this system (other than 1) which is not properly factorizable *in this system*. The numbers 5, 9, 13, 17, 21 are all pseudo-primes, and the first number in the series which is not a pseudo-prime is 25. It is true that every number in the system is either a pseudo-prime or can be factorized into pseudo-primes, and this can be proved in just the same way as before. But it is *not* true that the factorization is unique; for example, the number 693 can be factorized both as 9×77 and as 21×33, and the four numbers 9, 21, 33, 77 are all pseudo-primes. Of course, we know quite well that these numbers factorize further outside the system; but the point of the example lies in the light which it throws on the logical structure of any proof of the uniqueness of factorization. Such a proof *cannot* be based solely on the definition of a prime and on multiplicative operations. It must make use somewhere of addition or subtraction, for otherwise it would apply to this system of numbers too. If we examine the proof of the fundamental theorem given above we see that one operation of subtraction was used, namely in forming the number $n-pp'$.

The fundamental theorem of arithmetic exhibits the structure of the natural numbers in relation to the operation of multiplication. It shows us that the primes are the elements out of which all the natural numbers can be built up by multiplication in every possible way; moreover, when we carry out all these possible multiplications, the same number never arises in two different forms. It now becomes clear why it would have been inconvenient to classify the number 1 as a prime. Had we done so, we should have had to make an exception of it when stating that factorization into primes is possible in only one way; for obviously additional factors 1 can be introduced into any product without altering its value.

5. *Consequences of the fundamental theorem*

The fundamental theorem of arithmetic, which was proved in the last section, states that any natural number can be expressed as a product of primes in one and only one way, provided we admit products of one factor only to represent the primes themselves, and an empty product to represent the number 1.

If the factorization of a number into primes is known, then various questions concerning that number can be answered at once. In the first place, one can enumerate all the divisors of the number. Let us first see how this is done in a particular case. We take the same numerical example as before:

$$666 = 2 \times 3 \times 3 \times 37.$$

A divisor of this number is a number d such that

$$666 = dd',$$

where d' is another natural number. By the fundamental theorem of arithmetic, the factorizations of d and d' into primes must be such that when they are multiplied together the result is the product $2 \times 3 \times 3 \times 37$. So d must be the product of some of the primes 2, 3, 3, 37, and d' must be the product of the others. (The convention that we made earlier, concerning an empty product having the value 1, continues to be of service, since it permits the inclusion, under the same form of words, of the extreme cases when d or d' is 1.) On carrying out the choice of primes in every possible way, we get all the divisors of 666, namely

$$1, 2, 3, 37, 2 \times 3, 2 \times 37, 3 \times 3, 3 \times 37, 2 \times 3 \times 3,$$
$$2 \times 3 \times 37, 3 \times 3 \times 37, 2 \times 3 \times 3 \times 37.$$

The situation is perfectly general, and all that is needed is an appropriate notation in which to give a simple description of it. Let n be any natural number greater than 1, and let the *distinct* primes in its factorization be p, q, r, ... Suppose the prime p occurs a times in the factorization of n, the prime q occurs b times, and so on. Then

$$n = p^a q^b \dots . \tag{1}$$

The divisors of n consist of all the products

$$p^\alpha q^\beta \dots ,$$

where the exponent α has the possible values 0, 1, ... , a; the exponent β has the values 0, 1, ... , b; and so on.* This is proved in exactly the same way as in the preceding example and again the proof depends on the fundamental theorem of arithmetic. In the example with $n = 666$, there are three distinct prime factors, namely 2, 3, 37, and their exponents are 1, 2, 1 respectively. So all the divisors of 666 are given by the formula

$$2^\alpha 3^\beta 37^\gamma,$$

where α is 0 or 1, β is 0 or 1 or 2, and γ is 0 or 1. When written out, one at a time, these are the divisors enumerated above.

We can count how many divisors a number n has by counting how many choices there are for the exponents α, β, γ, ... In the general case, when n has the representation (1), the exponent α can be any one of 0, 1, ... , a, and so the number of different possibilities for α is $a+1$. Similarly the number of possibilities for β is $b+1$, and so on. The choices of the various exponents α, β, ... are independent of one another, and all the choices give different divisors of n, by the uniqueness of prime factorization. Hence the total number of divisors is

$$(a+1)(b+1)\dots .$$

It is usual to denote the number of divisors of a number n (includ-

*It is to be understood, as usual, that any number raised to the power zero means 1.

ing 1 and n, as we have done above) by $d(n)$. With this notation, we have proved that if $n=p^a q^b \dots$, where p, q, $\dots$ are distinct primes, then

$$d(n)=(a+1)(b+1) \dots .$$

In the above example, the exponents are 1, 2, 1 and the number of divisors is $2\times 3\times 2=12$.

One may also consider the *sum* of the divisors of n, again including 1 and n. This is usually denoted by $\sigma(n)$. If the factorization of n into primes is as written in (1), then $\sigma(n)$ is given by

$$\sigma(n)=\{1+p+p^2+ \dots +p^a\}\{1+q+q^2+ \dots +q^b\} \dots .$$

For this expression, when multiplied out, is the sum of all possible products of the form $p^\alpha q^\beta \dots$, where α takes each of the values $0, 1, \dots , a$, and so on. These possible products constitute all the divisors of n. To use again the same numerical illustration, we have

$$\sigma(666)=(1+2)(1+3+3^2)(1+37)=3\times 13\times 38=1482,$$

as one may check by working out all the divisors and adding them. The arithmetical functions $d(n)$ and $\sigma(n)$, and another function $\phi(n)$ which we shall meet later, are tabulated up to $n=10{,}000$ in *Number-divisor Tables* (vol. VIII of the British Ass. Math. Tables, Cambridge, 1940).

The ancient Greeks attached some importance to *perfect* numbers, which they defined as numbers n with the property that the sum of the divisors of n, including 1 but excluding n, is equal to n itself. The simplest example is 6, where $1+2+3=6$. An alternative way of expressing the definition is to say that $\sigma(n)=2n$, since $\sigma(n)$ is the sum of all the divisors, including n itself. It was proved by Euclid (Book IX, Prop. 36) that if p is any prime for which $p+1$ is a power of 2, say $p+1=2^k$, then the number $2^{k-1}p$ is perfect. In fact, we see from the above formula for $\sigma(n)$ that

$$\sigma(2^{k-1}p)=\{1+2+2^2+ \dots +2^{k-1}\}\{1+p\}.$$

Now

$$1+2+2^2+ \dots +2^{k-1}=2^k-1=p, \quad \text{and } 1+p=2^k,$$

whence $\sigma(n)=2n$ when $n=2^{k-1}p$. Euler, in a posthumous paper, supplemented Euclid's result by proving that every *even* perfect number is necessarily of Euclid's form. It is not known whether there are any odd perfect numbers, nor is it known whether the series of even perfect numbers continues indefinitely. The first five even perfect numbers are

$$6, \quad 28, \quad 496, \quad 8{,}128, \quad 33{,}550{,}336.$$

So far we have been considering the divisors of one number. We can also investigate the common divisors of two or more numbers. Any common divisor of two numbers m and n must be composed entirely of primes which occur in both m and n. If there are no such primes, then m and n have no common divisor apart from 1, and are said to be *relatively prime*, or *co-prime*. For example, the numbers

$$2829 = 3 \times 23 \times 41 \quad \text{and} \quad 6850 = 2 \times 5^2 \times 137$$

are relatively prime.

If m and n have common prime factors, we obtain the greatest common divisor or *highest common factor* (H.C.F.) of m and n by multiplying together the various common prime factors of m and n, each of these being taken to the highest power to which it divides *both m and n*. For example, if the two numbers are

$$3132 = 2^2 \times 3^3 \times 29 \quad \text{and} \quad 7200 = 2^5 \times 3^2 \times 5^2,$$

the H.C.F. is $2^2 \times 3^2$, or 36. It will be seen that the exponent of each prime in the H.C.F. is the lesser of the two exponents to which this prime occurs in the numbers m and n.

It is also obvious that the common divisors of m and n consist precisely of all divisors of their H.C.F. In making all these statements we are, of course, relying throughout upon the fundamental theorem of arithmetic.

A similar situation arises with the *common multiples* of two given numbers. There is a *least common multiple*, obtained by multiplying together all the primes which occur in *either m or n*, each of these being taken to the highest power to which it occurs in either number. Thus, for the two numbers written above, the L.C.M. is $2^5 \times 3^3 \times 5^2 \times 29$. The common multiples of two given numbers consist precisely of all multiples of their L.C.M.

These considerations extend easily to more than two numbers. But then it is important to note the two kinds of relative primality that are possible. Several numbers are said to be *relatively prime* if there is no number greater than 1 which divides all of them; they are said to be *relatively prime in pairs* if no two of them have a common factor greater than 1. The condition for the former is that there shall be no one prime occurring in all the numbers, and for the latter it is that there shall be no prime that occurs in any two of the numbers.

There are several simple theorems on divisibility which one is inclined to think of as obvious, but which are in fact only obvious in the light of the uniqueness of factorization into primes. For example, *if a number divides the product of two numbers and is relatively prime to one of them it must divide the other*. If *a* divides *bc* and is relatively prime to *b*, the prime factorization of *a* is contained in that of *bc* but has nothing in common with that of *b*, and is therefore contained in that of *c*.

6. *Euclid's algorithm*

In Prop. 2 of Book VII, Euclid gave a systematic process, or *algorithm*, for finding the highest common factor of two given numbers. This algorithm provides a different approach to questions of divisibility from that adopted in the last two sections, and we therefore begin again without assuming anything except the mere definition of divisibility.

Let *a* and *b* be two given natural numbers, and suppose that $a>b$. We propose to investigate the common divisors of *a* and *b*. If *a* is divisible by *b*, then the common divisors of *a* and *b* consist simply of all divisors of *b*, and there is no more to be said. If *a* is not divisible by *b*, we can express *a* as a multiple of *b* together with a remainder less than *b*, that is

$$a=qb+c, \quad \text{where} \quad c<b. \tag{2}$$

This is the process of 'division with a remainder', and expresses the fact that *a*, not being a multiple of *b*, must occur somewhere between two consecutive multiples of *b*. If *a* comes between qb and $(q+1)b$, then

$$a=qb+c, \quad \text{where} \quad 0<c<b.$$

It follows from the equation (2) that any common divisor of *b* and *c* is also a divisor of *a*. Moreover, any common divisor of *a* and *b* is also a divisor of *c*, since $c=a-qb$. It follows that the common divisors of *a* and *b*, whatever they may be, are the same as the common divisors of *b* and *c*. The problem of finding the common divisors of *a* and *b* is reduced to the same problem for the numbers *b* and *c*, which are respectively less than *a* and *b*.

The essence of the algorithm lies in the repetition of this argument. If b is divisible by c, the common divisors of b and c consist of all divisors of c. If not, we express b as

$$b = rc + d, \quad \text{where} \quad d < c. \tag{3}$$

Again, the common divisors of b and c are the same as those of c and d.

The process goes on until it terminates, and this can only happen when exact divisibility occurs, that is, when we come to a number in the sequence a, b, c, ... , which is a divisor of the preceding number. It is plain that the process must terminate, for the decreasing sequence a, b, c, ... of natural numbers cannot go on for ever.

Let us suppose, for the sake of definiteness, that the process terminates when we reach the number h, which is a divisor of the preceding number g. Then the last two equations of the series (2), (3), ... are

$$f = vg + h, \tag{4}$$

$$g = wh. \tag{5}$$

The common divisors of a and b are the same as those of b and c, or of c and d, and so on until we reach g and h. Since h divides g, the common divisors of g and h consist simply of all divisors of h. The number h can be identified as being the last remainder in Euclid's algorithm before exact divisibility occurs, i.e. the last non-zero remainder.

We have therefore proved that *the common divisors of two given natural numbers a and b consist of all divisors of a certain number h (the H.C.F. of a and b), and this number is the last non-zero remainder when Euclid's algorithm is applied to a and b.*

As a numerical illustration, take the numbers 3132 and 7200 which were used in §5. The algorithm runs as follows:

$$7200 = 2 \times 3132 + 936,$$
$$3132 = 3 \times 936 + 324,$$
$$936 = 2 \times 324 + 288,$$
$$324 = 1 \times 288 + 36,$$
$$288 = 8 \times 36;$$

and the H.C.F. is 36, the last remainder. It is often possible to shorten the working a little by using a negative remainder whenever this is numerically less than the corresponding positive remainder. In the above example, the last three steps could be replaced by

$$936 = 3 \times 324 - 36,$$
$$324 = 9 \times 36.$$

The reason why it is permissible to use negative remainders is that the argument that was applied to the equation (2) would be equally valid if that equation were $a = qb - c$ instead of $a = qb + c$.

Two numbers are said to be relatively prime* if they have no common divisor except 1, or in other words if their H.C.F. is 1. This will be the case if and only if the last remainder, when Euclid's algorithm is applied to the two numbers, is 1.

7. *Another proof of the fundamental theorem*

We shall now use Euclid's algorithm to give another proof of the fundamental theorem of arithmetic, independent of that given in §4.

We begin with a very simple remark, which may be thought to be too obvious to be worth making. Let a, b, n be any natural numbers. *The highest common factor of na and nb is n times the highest common factor of a and b.* However obvious this may seem, the reader will find that it is not easy to give a proof of it without using either Euclid's algorithm or the fundamental theorem of arithmetic.

In fact the result follows at once from Euclid's algorithm. We can suppose $a > b$. If we divide na by nb, the quotient is the same as before (namely q) and the remainder is nc instead of c. The equation (2) is replaced by

$$na = q \cdot nb + nc.$$

The same applies to the later equations; they are all simply multiplied throughout by n. Finally, the last remainder, giving the H.C.F. of na and nb, is nh, where h is the H.C.F. of a and b.

We apply this simple fact to prove the following theorem, often called Euclid's theorem, since it occurs as Prop. 30 of Book VII. *If a prime divides the product of two numbers, it must divide one of the numbers* (or possibly both of them). Suppose the prime p divides the product na of two numbers, and does not divide a. The only factors of p are 1 and p, and therefore the only common

---

*This is, of course, the same definition as in §5, but is repeated here because the present treatment is independent of that given previously.

factor of p and a is 1. Hence, by the theorem just proved, the H.C.F. of np and na is n. Now p divides np obviously, and divides na by hypothesis. Hence p is a common factor of np and na, and so is a factor of n, since we know that every common factor of two numbers is necessarily a factor of their H.C.F. We have therefore proved that if p divides na, and does not divide a, it must divide n; and this is Euclid's theorem.

The uniqueness of factorization into primes now follows. For suppose a number n has two factorizations, say

$$n=pqr \ldots =p'q'r' \ldots ,$$

where all the numbers $p, q, r, \ldots , p', q', r', \ldots$ are primes. Since p divides the product $p'(q'r' \ldots)$ it must divide either p' or $q'r' \ldots$. If p divides p' then $p=p'$ since both numbers are primes. If p divides $q'r' \ldots$ we repeat the argument, and ultimately reach the conclusion that p must equal one of the primes $p', q', r', \ldots$. We can cancel the common prime p from the two representations, and start again with one of those left, say q. Eventually it follows that all the primes on the left are the same as those on the right, and the two representations are the same.

This is the alternative proof of the uniqueness of factorization into primes, which was referred to in §4. It has the merit of resting on a general theory (that of Euclid's algorithm) rather than on a special device such as that used in §4. On the other hand, it is longer and less direct.

8. A property of the H.C.F

From Euclid's algorithm one can deduce a remarkable property of the H.C.F., which is not at all apparent from the original construction for the H.C.F. by factorization into primes (§5). The property is that *the highest common factor h of two natural numbers a and b is representable as the difference between a multiple of a and a multiple of b, that is*

$$h=ax-by$$

where x and y are natural numbers.

Since a and b are both multiples of h, any number of the form $ax-by$ is necessarily a multiple of h; and what the result asserts is that there are some values of x and y for which $ax-by$ is actually equal to h.

Before giving the proof, it is convenient to note some properties of numbers representable as $ax-by$. In the first place, a number so representable can also be represented as $by'-ax'$, where x' and y' are natural numbers. For the two expressions will be equal if

$$a(x+x')=b(y+y');$$

and this can be ensured by taking any number m and defining x' and y' by

$$x+x'=mb, \quad y+y'=ma.$$

These numbers x' and y' will be natural numbers provided m is sufficiently large, so that $mb>x$ and $ma>y$. If x' and y' are defined in this way, then $ax-by=by'-ax'$.

We say that a number is *linearly dependent on a and b* if it is representable as $ax-by$. The result just proved shows that linear dependence on a and b is not affected by interchanging a and b.

There are two further simple facts about linear dependence. If a number is linearly dependent on a and b, then so is any multiple of that number, for

$$k(ax-by)=a.kx-b.ky.$$

Also the sum of two numbers that are each linearly dependent on a and b is itself linearly dependent on a and b, since

$$(ax_1-by_1)+(ax_2-by_2)=a(x_1+x_2)-b(y_1+y_2).$$

The same applies to the difference of two numbers; to see this, write the second number as $by_2'-ax_2'$, in accordance with the earlier remark, before subtracting it. Then we get

$$(ax_1-by_1)-(by_2'-ax_2')=a(x_1+x_2')-b(y_1+y_2').$$

So the property of linear dependence on a and b is preserved by addition and subtraction, and by multiplication by any number.

We now examine the steps in Euclid's algorithm, in the light of this concept. The numbers a and b themselves are certainly linearly dependent on a and b, since

$$a=a(b+1)-b(a), \quad b=a(b)-b(a-1).$$

The first equation of the algorithm was

$$a=qb+c.$$

Since b is linearly dependent on a and b, so is qb, and since a is also linearly dependent on a and b, so is $a-qb$, that is c. Now the next equation of the algorithm allows us to deduce in the same way that d is linearly dependent on a and b, and so on until we come to the last remainder, which is h. This proves that h is linearly dependent on a and b, as asserted.

As an illustration, take the same example as was used in §6, namely $a=7200$ and $b=3132$. We work through the equations one at a time, using them to express each remainder in terms of a and b. The first equation was

$$7200 = 2 \times 3132 + 936,$$

which tells us that

$$936 = a - 2b.$$

The second equation was $3132 = 3 \times 936 + 324$, which gives

$$324 = b - 3(a - 2b) = 7b - 3a.$$

The third equation was

$$936 = 2 \times 324 + 288,$$

which gives

$$288 = (a - 2b) - 2(7b - 3a) = 7a - 16b.$$

The fourth equation was

$$324 = 1 \times 288 + 36,$$

which gives

$$36 = (7b - 3a) - (7a - 16b) = 23b - 10a.$$

This expresses the highest common factor, 36, as the difference of two multiples of the numbers a and b. If one prefers an expression in which the multiple of a comes first, this can be obtained by arguing that

$$23b - 10a = (M - 10)a - (N - 23)b,$$

provided that $Ma = Nb$. Since a and b have the common factor 36, this factor can be removed from both of them, and the condition on M and N becomes $200M = 87N$. The simplest choice for M and N is $M = 87$, $N = 200$, which on substitution gives

$$36 = 77a - 177b.$$

Returning to the general theory, we can express the result in another form. Suppose a, b, n are given natural numbers, and it is desired to find natural numbers x and y such that

$$ax - by = n. \qquad (6)$$

Such an equation is called an *indeterminate* equation since it does not determine x and y completely, or a *Diophantine* equation after Diophantus of Alexandria (third century A.D.), who wrote a famous treatise on arithmetic. The equation (6) cannot be soluble unless n is a multiple of the highest common factor h of a and b; for this highest common factor divides $ax - by$, whatever values x and y may have. Now suppose that n is a multiple of h, say, $n = mh$. Then we can solve the equation; for all we have to do is first solve the equation

$$ax_1 - by_1 = h,$$

as we have seen how to do above, and then multiply throughout by m, getting the solution $x = mx_1$, $y = my_1$ for the equation (6). Hence *the linear indeterminate equation* (6) *is soluble in natural numbers x, y if and only if n is a multiple of h*. In particular, if a and b are relatively prime, so that $h = 1$, the equation is soluble whatever value n may have.

As regards the linear indeterminate equation

$$ax + by = n,$$

we have found the condition for it to be soluble, not in natural numbers, but in integers of opposite signs: one positive and one negative. The question of when this equation is soluble in natural numbers is a more difficult one, and one that cannot well be completely answered in any simple way. Certainly n must be a multiple of h, but also n must not be too small in relation to a and b. It can be proved quite easily that the equation is soluble in natural numbers if n is a multiple of h and $n > ab$.

9. Factorizing a number

The obvious way of factorizing a number is to test whether it is divisible by 2 or by 3 or by 5, and so on, using the series of primes. If a number N is not divisible by any prime up to $\sqrt{N}$, it must be itself a prime; for any composite number has at least two prime factors, and they cannot both be greater than $\sqrt{N}$.

The process is a very laborious one if the number is at all large, and for this reason factor tables have been computed. The most extensive one which is generally accessible is that of D. N. Lehmer (Carnegie Institute, Washington, Pub. No. 105. 1909; reprinted by Hafner Press, New York, 1956), which gives the least prime factor of each number up to 10,000,000. When the least prime factor of a particular number is known, this can be divided out, and repetition of the process gives eventually the complete factorization of the number into primes.

Several mathematicians, among them Fermat and Gauss, have invented methods for reducing the amount of trial that is necessary to factorize a large number. Most of these involve more knowledge of number-theory than we can postulate at this stage; but there is one method of Fermat which is in principle extremely simple and can be explained in a few words.

Let N be the given number, and let m be the least number for which $m^2 > N$. Form the numbers

$$m^2 - N, \quad (m+1)^2 - N, \quad (m+2)^2 - N, \ldots . \tag{7}$$

When one of these is reached which is a perfect square, we get $x^2 - N = y^2$, and consequently $N = x^2 - y^2 = (x-y)(x+y)$. The calculation of the numbers (7) is facilitated by noting that their successive differences increase at a constant rate. The identification of one of them as a perfect square is most easily made by using Barlow's Table of Squares. The method is particularly successful if the number N has a factorization in which the two factors are of about the same magnitude, since then y is small. If N is itself a prime, the process goes on until we reach the solution provided by $x + y = N$, $x - y = 1$.

As an illustration, take $N = 9271$. This comes between 96^2 and 97^2, so that $m = 97$. The first number in the series (7) is $97^2 - 9271 = 138$. The subsequent ones are obtained by adding successively $2m + 1$, then $2m + 3$, and so on, that is, 195, 197, and so on. This gives the series

$$138, 333, 530, 729, 930, \ldots .$$

The fourth of these is a perfect square, namely 27^2, and we get

$$9271 = 100^2 - 27^2 = 127 \times 73.$$

An interesting algorithm for factorization has been discovered recently by Captain N. A. Draim, u.s.n. In this, the result of each trial division is used to modify the number in preparation for the next division. There are several forms of the algorithm, but perhaps the simplest is that in which the successive divisors are the odd numbers 3, 5, 7, 9, ... , whether prime or not. To explain the rules, we work a numerical example, say $N=4511$. The first step is to divide by 3, the quotient being 1503 and the remainder 2:

$$4511 = 3 \times 1503 + 2.$$

The next step is to subtract twice the quotient from the given number, and then add the remainder:

$$4511 - 2 \times 1503 - 1505, \quad 1505 + 2 = 1507.$$

The last number is the one which is to be divided by the next odd number, 5:

$$1507 = 5 \times 301 + 2.$$

The next step is to subtract twice the quotient from the first derived number on the previous line (1505 in this case), and then add the remainder from the last line:

$$1505 - 2 \times 301 = 903, \quad 903 + 2 = 905.$$

This is the number which is to be divided by the next odd number, 7. Now we can continue in exactly the same way, and no further explanation will be needed:

$$905 = 7 \times 129 + 2,$$
$$903 - 2 \times 129 = 645, \quad 645 + 2 = 647,$$
$$647 = 9 \times 71 + 8,$$
$$645 - 2 \times 71 = 503, \quad 503 + 8 = 511,$$
$$511 = 11 \times 46 + 5,$$
$$503 - 2 \times 46 = 411, \quad 411 + 5 = 416,$$
$$416 = 13 \times 32 + 0.$$

We have reached a zero remainder, and the algorithm tells us that 13 is a factor of the given number 4511. The complementary factor is found by carrying out the first half of the next step:

$$411 - 2 \times 32 = 347.$$

In fact $4511 = 13 \times 347$, and as 347 is a prime the factorization is complete.

To justify the algorithm generally is a matter of elementary algebra. Let N_1 be the given number; the first step was to express N_1 as

$$N_1 = 3q_1 + r_1.$$

The next step was to form the numbers

$$M_2 = N_1 - 2q_1, \quad N_2 = M_2 + r_1.$$

The number N_2 was divided by 5:

$$N_2 = 5q_2 + r_2,$$

and the next step was to form the numbers

$$M_3 = M_2 - 2q_2, \quad N_3 = M_3 + r_2,$$

and so the process was continued. It can be deduced from these equations that

$$N_2 = 2N_1 - 5q_1,$$
$$N_3 = 3N_1 - 7q_1 - 7q_2,$$
$$N_4 = 4N_1 - 9q_1 - 9q_2 - 9q_3,$$

and so on. Hence N_2 is divisible by 5 if and only if $2N_1$ is divisible by 5, or N_1 divisible by 5. Again, N_3 is divisible by 7 if and only if $3N_1$ is divisible by 7, or N_1 divisible by 7, and so on. When we reach as divisor the least prime factor of N_1, exact divisibility occurs and there is a zero remainder.

The general equation analogous to those given above is

$$N_n = nN_1 - (2n+1)(q_1 + q_2 + \ldots + q_{n-1}). \tag{8}$$

The general equation for M_n is found to be

$$M_n = N_1 - 2(q_1 + q_2 + \ldots + q_{n-1}). \tag{9}$$

If $2n+1$ is a factor of the given number N_1, then N_n is exactly divisible by $2n+1$, and

$$N_n = (2n+1)q_n,$$

whence

$$nN_1 = (2n+1)(q_1 + q_2 + \ldots + q_n),$$

by (8). Under these circumstances, we have, by (9),

$$M_{n+1} = N_1 - 2(q_1 + q_2 + \ldots + q_n)$$
$$= N_1 - 2\left(\frac{n}{2n+1}\right)N_1 = \frac{N_1}{2n+1}.$$

Thus the complementary factor to the factor $2n+1$ is M_{n+1}, as stated in the example.

In the numerical example worked out above, the numbers N_1, N_2, ... decrease steadily. This is always the case at the beginning of the algorithm, but may not be so later. However, it appears that

the later numbers are always considerably less than the original
number.

10. The series of primes

Although the notion of a prime is a very natural and obvious one,
questions concerning the primes are often very difficult, and many
such questions are quite unanswerable in the present state of
mathematical knowledge. We conclude this chapter by mentioning
briefly some results and conjectures about the primes.

In §3 we gave Euclid's proof that there are infinitely many
primes. The same argument will also serve to prove that there are infinitely many primes of certain specified forms. Since every prime
after 2 is odd, each of them falls into one of the two progressions

(a) 1, 5, 9, 13, 17, 21, 25, ... ,
(b) 3, 7, 11, 15, 19, 23, 27, ... ;

the progression (a) consisting of all numbers of the form $4x+1$,
and the progression (b) of all numbers of the form $4x-1$ (or
$4x+3$, which comes to the same thing). We first prove that *there
are infinitely many primes in the progression* (b). Let the primes in
(b) be enumerated as $q_1, q_2, ...$, beginning with $q_1=3$. Consider
the number N defined by

$$N=4(q_1q_2 \cdots q_n)-1.$$

This is itself a number of the form $4x-1$. Not every prime factor
of N can be of the form $4x+1$, because any product of numbers
which are all of the form $4x+1$ is itself of that form, e.g.

$$(4x+1)(4y+1)=4(4xy+x+y)+1.$$

Hence the number N has *some* prime factor of the form $4x-1$.
This cannot be any of the primes $q_1, q_2, ... , q_n$, since N leaves the
remainder -1 when divided by any of them. Thus there exists a
prime in the series (b) which is different from any of $q_1, q_2, ... , q_n$;
and this proves the proposition.

The same argument cannot be used to prove that there are
infinitely many primes in the series (a), because if we construct a
number of the form $4x+1$ it does not follow that this number will
necessarily have a prime factor of that form. However, another
argument can be used. Let the primes in the series (a) be enumerated as $r_1, r_2, ...$, and consider the number M defined by

$$M=(r_1r_2 \cdots r_n)^2+1.$$

We shall see later (III.3) that any number of the form a^2+1 has a prime factor of the form $4x+1$, and is indeed entirely composed of such primes, together possibly with the prime 2. Since M is obviously not divisible by any of the primes $r_1, r_2, \ldots, r_n$, it follows as before that *there are infinitely many primes in the progression* (a).

A similar situation arises with the two progressions $6x+1$ and $6x-1$. These progressions exhaust all numbers that are not divisible by 2 or 3, and therefore every prime after 3 falls in one of these two progressions. One can prove by methods similar to those used above that there are infinitely many primes in each of them. But such methods cannot cope with the general arithmetical progression. Such a progression consists of all numbers $ax+b$, where a and b are fixed and $x=0, 1, 2, \ldots$, that is, the numbers

$$b, \quad b+a, \quad b+2a, \quad \ldots .$$

If a and b have a common factor, every number of the progression has this factor, and so is not a prime (apart from possibly the first number b). We must therefore suppose that a and b are relatively prime. It then seems plausible that the progression will contain infinitely many primes, i.e. that *if a and b are relatively prime, there are infinitely many primes of the form $ax+b$.*

Legendre seems to have been the first to realize the importance of this proposition. At one time he thought he had a proof, but this turned out to be fallacious. The first proof was given by Dirichlet in an important memoir which appeared in 1837. This proof used analytical methods (functions of a continuous variable, limits, and infinite series), and was the first really important application of such methods to the theory of numbers. It opened up completely new lines of development; the ideas underlying Dirichlet's argument are of a very general character and have been fundamental for much subsequent work applying analytical methods to the theory of numbers.

Not much is known about other forms which represent infinitely many primes. It is conjectured, for instance, that there are infinitely many primes of the form x^2+1, the first few being

$$2, 5, 17, 37, 101, 197, 257, \ldots .$$

But not the slightest progress has been made towards proving this, and the question seems hopelessly difficult. Dirichlet did succeed, however, in proving that any quadratic form in two variables, that is, any form $ax^2+bxy+cy^2$, in which a, b, c are relatively prime, represents infinitely many primes.

A question which has been deeply investigated in modern times is that of the frequency of occurrence of the primes, in other words the question of how many primes there are among the numbers $1, 2, \ldots, X$ when X is large. This number, which depends of course on X, is usually denoted by $\pi(X)$. The first conjecture about the magnitude of $\pi(X)$ as a function of X seems to have been made independently by Legendre and Gauss about 1800. It was that $\pi(X)$ is approximately $\dfrac{X}{\log X}$. Here $\log X$ denotes the natural (so-called Napierian) logarithm of X, that is, the logarithm of X to the base e. The conjecture seems to have been based on numerical evidence. For example, when X is 1,000,000 it is found that $\pi(1,000,000) = 78,498$, whereas the value of $X/\log X$ (to the nearest integer) is 72,382, the ratio being 1.084 … . Numerical evidence of this kind may, of course, be quite misleading. But here the result suggested is true, in the sense that the ratio of $\pi(X)$ to $X/\log X$ tends to the limit 1 as X tends to infinity. This is the famous Prime Number Theorem, first proved by Hadamard and de la Vallée Poussin independently in 1896, by the use of new and powerful analytical methods.

It is impossible to give an account here of the many other results which have been proved concerning the distribution of the primes. Those proved in the nineteenth century were mostly in the nature of imperfect approaches towards the Prime Number Theorem; those of the twentieth century included various refinements of that theorem. There is one recent event to which, however, reference should be made. We have already said that the proof of Dirichlet's Theorem on primes in arithmetical progressions and the proof of the Prime Number Theorem were analytical, and made use of methods which cannot be said to belong properly to the theory of numbers. The propositions themselves relate entirely to the natural numbers, and it seems reasonable that they should be provable without the intervention of such foreign ideas. The search for 'elementary' proofs of these two theorems was unsuccessful until fairly recently. In 1948 A. Selberg found the first elementary proof of Dirichlet's Theorem, and with the help of P. Erdös he found the first elementary proof of the Prime Number Theorem. An 'elementary' proof, in this connection, means a proof which operates only with natural numbers. Such a proof is not necessarily simple, and indeed both the proofs in question are distinctly difficult.

Finally, we may mention the famous problem concerning primes which was propounded by Goldbach in a letter to Euler in 1742. Goldbach suggested (in a slightly different wording) that every even number from 6 onwards is representable as the sum of two primes other than 2, e.g.

$$6=3+3, \ 8=3+5, \ 10=3+7=5+5, \ 12=5+7, \ \dots .$$

Any problem like this which relates to *additive* properties of primes is necessarily difficult, since the definition of a prime and the natural properties of primes are all expressed in terms of multiplication. An important contribution to the subject was made by Hardy and Littlewood in 1923, but it was not until 1930 that anything was rigorously proved that could be considered as even a remote approach towards a solution of Goldbach's problem. In that year the Russian mathematician Schnirelmann proved that there is some number N such that *every number from some point onwards is representable as the sum of at most N primes.* A much nearer approach was made by Vinogradov in 1937. He proved, by analytical methods of extreme subtlety, that *every odd number from some point onwards is representable as the sum of three primes.* This was the starting point of much new work on the additive theory of primes, in the course of which many problems have been solved which would have been quite beyond the scope of any pre-Vinogradov methods. A recent result in connection with Goldbach's problem is that every sufficiently large even number is representable as the sum of two numbers, one of which is a prime and the other of which has at most two prime factors.

NOTES

§1. The main difficulty in giving any account of the laws of arithmetic, such as that given here, lies in deciding which of the various concepts should come first. There are several possible arrangements, and it seems to be a matter of taste which one prefers.

It is no part of our purpose to analyse further the concepts and laws of arithmetic. We take the commonsense (or naïve) view that we all 'know' the natural numbers, and are satisfied of the validity of the laws of arithmetic and of the principle of induction. The reader who is interested in the foundations of mathematics may consult Bertrand Russell, *Introduction to Mathematical Philosophy* (Allen and Unwin, London), or M. Black, *The Nature of Mathematics* (Harcourt, Brace, New York).

Russell defines the natural numbers by selecting them from numbers of a more general kind. These more general numbers are the (finite or infinite) cardinal numbers, which are defined by means of the general notions of 'class' and 'one-to-one correspondence.' The selection is made by defining the natural numbers as those which possess all inductive properties (Russell, *loc. cit.*, p. 27). But whether it is reasonable to base the theory of the natural numbers on such a vague and unsatisfactory concept as that of a class is a matter of opinion. '*Dolus latet in universalibus,*' as Dr Johnson remarked.

§2. The objection to using the principle of induction as a definition of the natural numbers is that it involves references to 'any proposition about a natural number n'. It seems plain that the 'propositions' envisaged here must be statements which are significant when made about natural numbers. It is not clear how this significance can be tested or appreciated except by one who already knows the natural numbers.

§4. I am not aware of having seen this proof of the uniqueness of prime factorization elsewhere, but it is unlikely that it is new. For other direct proofs, see Mathews, p. 2, or Hardy and Wright, p. 21.*

§5. It has been shown by computer searches that there is no odd perfect number less than 10^{200}. If an odd perfect number exists, it has at least eight distinct prime divisors, of which the largest exceeds 100,110. For references and other information on perfect numbers, see Guy, sections A.3 and B.1.

§6. A critical reader may notice that in two places in this section I have used principles that were not explicitly stated in §§ 1 and 2. In each place, a proof by induction could have been given, but to have done so would have distracted the reader's attention from the main issues.

The question of the *length* of Euclid's algorithm is discussed in Uspensky and Heaslet, ch. 3, and D. E. Knuth's *The Art of Computer Programming*, vol. II: *Seminumerical Algorithms* (Addison Wesley, Reading, Mass., 2nd ed., 1981) section 4.5.3.

§9. For an account of early methods of factoring, see Dickson's *History*, vol. I, ch. 14. For a discussion of modern methods see the article by Richard K. Guy, 'How to factor a number', *Congressus Numerantium XVI Proc. 5th Manitoba Conf. Numer. Math.*, Winnipeg, 1975, 49–89. A recent method is given by L. Adleman and F. T. Leighton, 'An $O(n^{1/10\cdot89})$ primality testing algorithm', *Math. Comput.*, 36 (1981), 261–6. It is doubtful whether D. N. Lehmer's tables will ever be extended, since with them and a pocket calculator one can easily check whether a twelve-digit number is a prime. For Draim's algorithm, see *Mathematics Magazine*, 25 (1952), 191–4.

*Particulars of books referred to by their authors' names will be found in the Bibliography.

§10. An excellent account of the distribution of primes is given in A. E. Ingham, *The Distribution of Prime Numbers* (Cambridge Tracts, no. 30, 1932; reprinted by Hafner Press, New York, 1971). For a more recent and extensive account see H. Davenport, *Multiplicative Number Theory*, 2nd ed. (Springer, 1980). H. Iwaniec (*Inventiones Math.*, 47 (1978), 171–88) has shown that for infinitely many n the number $n^2 + 1$ is either a prime or a product of two primes. Dirichlet's proof of his theorem (with a modification due to Mertens) is given as an appendix to Dickson's *Modern Elementary Theory of Numbers*. An elementary proof of the Prime Number Theorem is given in ch. 22 of Hardy and Wright. An elementary proof of the asymptotic formula for the number of primes in an arithmetic progression is given in Gelfond and Linnik, ch. 3. For a survey of early work on Goldbach's problem, see James, *Bull. American Math. Soc.*, 55 (1949), 246–60. For a proof of Chen's theorem that every sufficiently large even integer can be represented as $p + P_2$, where p is a prime, and P_2 is either a prime or the product of two primes, see ch. 11 of *Sieve Methods* by H. Halberstam and H. E. Richert (Academic Press, London, 1974). For a proof of Vinogradov's result, see T. Estermann, *Introduction to Modern Prime Number Theory* (Cambridge Tracts, no. 41, 1952) or H. Davenport, *Multiplicative Number Theory*, 2nd ed. (Springer, 1980).

II

CONGRUENCES

1. The congruence notation

It often happens that for the purposes of a particular calculation, two numbers which differ by a multiple of some fixed number are equivalent, in the sense that they produce the same result. For example, the value of $(-1)^n$ depends only on whether n is odd or even, so that two values of n which differ by a multiple of 2 give the same result. Or again, if we are concerned only with the last digit of a number, then for that purpose two numbers which differ by a multiple of 10 are effectively the same.

The congruence notation, introduced by Gauss, serves to express in a convenient form the fact that two integers a and b differ by a multiple of a fixed natural number m. We say that *a is congruent to b with respect to the modulus m*, or, in symbols,

$$a \equiv b \pmod{m}.$$

The meaning of this, then, is simply that $a-b$ is divisible by m. The notation facilitates calculations in which numbers differing by a multiple of m are effectively the same, by stressing the analogy between congruence and equality. Congruence, in fact, means 'equality except for the addition of some multiple of m'.

A few examples of valid congruences are:

$$63 \equiv 0 \pmod{3}, \quad 7 \equiv -1 \pmod{8}, \quad 5^2 \equiv -1 \pmod{13}.$$

A congruence to the modulus 1 is always valid, whatever the two numbers may be, since every number is a multiple of 1. Two numbers are congruent with respect to the modulus 2 if they are of the same parity, that is, both even or both odd.

Two congruences can be added, subtracted, or multiplied together, in just the same way as two equations, provided all the congruences have the same modulus. If

$$a \equiv \alpha \ (\mathrm{mod}\ m) \quad \text{and} \quad b \equiv \beta \ (\mathrm{mod}\ m)$$

then

$$a+b \equiv \alpha+\beta \ (\mathrm{mod}\ m),$$
$$a-b \equiv \alpha-\beta \ (\mathrm{mod}\ m),$$
$$ab \equiv \alpha\beta \quad (\mathrm{mod}\ m).$$

The first two of these statements are immediate; for example $(a+b)-(\alpha+\beta)$ is a multiple of m because $a-\alpha$ and $b-\beta$ are both multiples of m. The third is not quite so immediate and is best proved in two steps. First $ab \equiv \alpha b$ because $ab-\alpha b = (a-\alpha)b$, and $a-\alpha$ is a multiple of m. Next, $\alpha b \equiv \alpha\beta$, for a similar reason. Hence $ab \equiv \alpha\beta \ (\mathrm{mod}\ m)$.

A congruence can always be multiplied throughout by any integer: if $a \equiv \alpha \ (\mathrm{mod}\ m)$ then $ka \equiv k\alpha \ (\mathrm{mod}\ m)$. Indeed this is a special case of the third result above, where b and β are both k. But it is not always legitimate to cancel a factor from a congruence. For example

$$42 \equiv 12 \ (\mathrm{mod}\ 10),$$

but it is not permissible to cancel the factor 6 from the numbers 42 and 12, since this would give the false result $7 \equiv 2 \ (\mathrm{mod}\ 10)$. The reason is obvious: the first congruence states that $42-12$ is a multiple of 10, but this does not imply that $\frac{1}{6}(42-12)$ is a multiple of 10. The cancellation of a factor from a congruence *is* legitimate if the factor is *relatively prime to the modulus*. For let the given congruence be $ax \equiv ay \ (\mathrm{mod}\ m)$, where a is the factor to be cancelled, and we suppose that a is relatively prime to m. The congruence states that $a(x-y)$ is divisible by m, and it follows from the last proposition in I.5 that $x-y$ is divisible by m.

An illustration of the use of congruences is provided by the well-known rules for the divisibility of a number by 3 or 9 or 11. The usual representation of a number n by digits in the scale of 10 is really a representation of n in the form

$$n = a + 10b + 100c + \dots,$$

where $a, b, c, \dots$ are the digits of the number, read from right to left, so that a is the number of units, b the number of tens, and so on. Since $10 \equiv 1$ (mod 9), we have also $10^2 \equiv 1$ (mod 9), $10^3 \equiv 1$ (mod 9), and so on. Hence it follows from the above representation of n that

$$n \equiv a+b+c+ \ldots \text{ (mod 9).}$$

In other words, any number n differs from the sum of its digits by a multiple of 9, and in particular n is divisible by 9 if and only if the sum of its digits is divisible by 9. The same applies with 3 in place of 9 throughout.

The rule for 11 is based on the fact that $10 \equiv -1$ (mod 11), so that $10^2 \equiv +1$ (mod 11), $10^3 \equiv -1$ (mod 11), and so on. Hence

$$n \equiv a-b+c- \ldots \text{ (mod 11).}$$

It follows that n is divisible by 11 if and only if $a-b+c- \ldots$ is divisible by 11. For example, to test the divisibility of 9581 by 11 we form $1-8+5-9$, or -11. Since this is divisible by 11, so is 9581.

2. Linear congruences

It is obvious that every integer is congruent (mod m) to exactly one of the numbers

$$0, 1, 2, \ldots, m-1. \tag{1}$$

For we can express the integer in the form $qm+r$, where $0 \leqq r < m$, and then it is congruent to r (mod m). Obviously there are other sets of numbers, besides the set (1), which have the same property, e.g. any integer is congruent (mod 5) to exactly one of the numbers $0, 1, -1, 2, -2$. Any such set of numbers is said to constitute a *complete set of residues* to the modulus m. Another way of expressing the definition is to say that a complete set of residues (mod m) is any set of m numbers, no two of which are congruent to one another.

A *linear* congruence, by analogy with a linear equation in elementary algebra, means a congruence of the form

$$ax \equiv b \text{ (mod } m\text{).} \tag{2}$$

It is an important fact that any such congruence is soluble for x, provided that a is relatively prime to m. The simplest way of proving this is to observe that if x runs through the numbers of a complete set of residues, then the corresponding values of ax also constitute a complete set of residues. For there are m of these numbers, and no two of them are congruent, since $ax_1 \equiv ax_2$ (mod m) would involve $x_1 \equiv x_2$ (mod m), by the cancellation of the factor a (permissible since a is relatively prime to m). Since the numbers ax form a complete set of residues, there will be exactly one of them congruent to the given number b.

As an example, consider the congruence

$$3x \equiv 5 \text{ (mod 11)}.$$

If we give x the values 0, 1, 2, ... 10 (a complete set of residues to the modulus 11), $3x$ takes the values 0, 3, 6, ... , 30. These form another complete set of residues (mod 11), and in fact they are congruent respectively to

$$0, 3, 6, 9, 1, 4, 7, 10, 2, 5, 8.$$

The value 5 occurs when $x = 9$, and so $x = 9$ is a solution of the congruence. Naturally any number congruent to 9 (mod 11) will also satisfy the congruence; but nevertheless we say that the congruence has *one* solution, meaning that there is one solution in any complete set of residues. In other words, all solutions are mutually congruent. The same applies to the general congruence (2); such a congruence (provided a is relatively prime to m) is precisely equivalent to the congruence $x \equiv x_0$ (mod m), where x_0 is one particular solution.

There is another way of looking at the linear congruence (2). It is equivalent to the *equation* $ax = b + my$, or $ax - my = b$. We proved in I.8 that such a linear Diophantine equation is soluble for x and y if a and m are relatively prime, and that fact provides another proof of the solubility of the linear congruence. But the proof given above is simpler, and illustrates the advantages gained by using the congruence notation.

The fact that the congruence (2) has a unique solution, in the sense explained above, suggests that one may use this solution as an interpretation for the fraction $\dfrac{b}{a}$ to the modulus m. When we do this, we obtain an arithmetic (mod m) in which addition, subtraction and multiplication are always possible, and division is also possible provided that the divisor is relatively prime to m. In this arithmetic there are only a finite number of essentially distinct numbers, namely m of them, since two numbers which are mutually congruent (mod m) are treated as the same. If we take the modulus m to be 11, as an illustration, a few examples of 'arithmetic mod 11' are:

$$5+7 \equiv 1, \quad 5 \times 6 \equiv 8, \quad \tfrac{5}{3} \equiv 9 \equiv -2.$$

Any relation connecting integers or fractions in the ordinary sense remains true when interpreted in this arithmetic. For example, the relation

$$\tfrac{1}{2} + \tfrac{2}{3} = \tfrac{7}{6}$$

becomes (mod 11)

$$6+8 \equiv 3,$$

because the solution of $2x \equiv 1$ is $x \equiv 6$, that of $3x \equiv 2$ is $x \equiv 8$, and that of $6x \equiv 7$ is $x \equiv 3$. Naturally the interpretation given to a fraction depends on the modulus, for instance $\tfrac{2}{3} \equiv 8$ (mod 11), but $\tfrac{2}{3} \equiv 3$ (mod 7). The only limitation on such calculations is that just mentioned, namely that the denominator of any fraction must be relatively prime to the modulus. If the modulus is a prime (as in the above examples with 11), the limitation takes the very simple form that the denominator must not be congruent to 0 (mod m), and this is exactly analogous to the limitation in ordinary arithmetic that the denominator must not be equal to 0. We shall return to this point later (§7).

3. *Fermat's theorem*

The fact that there are only a finite number of essentially different numbers in arithmetic to a modulus m means that there are algebraic relations which are satisfied by *every* number in that arithmetic. There is nothing analogous to these relations in ordinary arithmetic.

Suppose we take any number x and consider its powers $x, x^2, x^3, \ldots$. Since there are only a finite number of possibilities for these to the modulus m, we must eventually come to one which we have met before, say $x^h \equiv x^k$ (mod m), where $k < h$. If x is relatively prime to m, the factor x^k can be cancelled, and it follows that $x^l \equiv 1$ (mod m), where $l = h - k$. Hence every number x which is relatively prime to m satisfies some congruence of this form. The *least* exponent l for which $x^l \equiv 1$ (mod m) will be called *the order of x to the modulus m*. If x is 1, its order is obviously 1. To illustrate the definition, let us calculate the orders of a few numbers to the modulus 11. The powers of 2, taken to the modulus 11, are

$$2, 4, 8, 5, 10, 9, 7, 3, 6, 1, 2, 4, \ldots .$$

Each one is twice the preceding one, with 11 or a multiple of 11 subtracted where necessary to make the result less than 11. The first power of 2 which is $\equiv 1$ is 2^{10}, and so the order of 2 (mod 11) is 10. As another example, take the powers of 3:

$$3, 9, 5, 4, 1, 3, 9, \ldots .$$

The first power of 3 which is $\equiv 1$ is 3^5, so the order of 3 (mod 11) is 5. It will be found that the order of 4 is again 5, and so also is that of 5.

It will be seen that the successive powers of x are periodic; when we have reached the first number l for which $x^l \equiv 1$, then $x^{l+1} \equiv x$ and the previous cycle is repeated. It is plain that $x^n \equiv 1$ *(mod m)* *if and only if n is a multiple of the order of x*. In the last example, $3^n \equiv 1$ (mod 11) if and only if n is a multiple of 5. This remains valid if n is 0 (since $3^0 = 1$), and it remains valid also for negative exponents, provided 3^{-n}, or $1/3^n$, is interpreted as a fraction (mod 11) in the way explained in §2. In fact, the negative powers of 3 (mod 11) are obtained by prolonging the series backwards, and the table of powers of 3 to the modulus 11 is

$n =$	...	-3	-2	-1	0	1	2	3	4	5	6	...
$3^n \equiv$	...	9	5	4	1	3	9	5	4	1	3	... .

Fermat discovered that if the modulus is a prime, say p, then every integer x not congruent to 0 satisfies

$$x^{p-1} \equiv 1 \pmod{p}. \tag{3}$$

In view of what we have seen above, this is equivalent to saying that the order of any number is a divisor of $p-1$. The result (3) was mentioned by Fermat in a letter to Frénicle de Bessy of 18 October 1640, in which he also stated that he had a proof. But as with most of Fermat's discoveries, the proof was not published or preserved. The first known proof seems to have been given by Leibniz (1646–1716). He proved that $x^p \equiv x$ (mod p), which is equivalent to (3), by writing x as a sum $1+1+ \ldots +1$ of x units (assuming x positive), and then expanding $(1+1+ \ldots +1)^p$ by the multinomial theorem. The terms $1^p + 1^p + \ldots + 1^p$ give x, and the coefficients of all the other terms are easily proved to be divisible by p.

Quite a different proof was given by Ivory in 1806. If $x \not\equiv 0$ (mod p), the integers

$$x, 2x, 3x, \ldots, (p-1)x$$

are congruent (in some order) to the numbers

$$1, 2, 3, \ldots, p-1.$$

In fact, each of these sets constitutes a complete set of residues except that 0 has been omitted from each. Since the two sets are congruent, their products are congruent, and so

$$(x)(2x)(3x) \ldots ((p-1)x) \equiv (1)(2)(3) \ldots (p-1) \pmod{p}.$$

Cancelling the factors 2, 3, ... , $p-1$, as is permissible, we obtain (3).

One merit of this proof is that it can be extended so as to apply to the more general case when the modulus is no longer a prime. The generalization of the result (3) to any modulus was first given by Euler in 1760. To formulate it, we must begin by considering how many numbers in the set 0, 1, 2, ... , $m-1$ are relatively prime to m. Denote this number by $\phi(m)$. When m is a prime, all the numbers in the set except 0 are relatively prime to m, so that $\phi(p) = p-1$ for any prime p. Euler's generalization of Fermat's theorem is that for any modulus m,

$$x^{\phi(m)} \equiv 1 \pmod{m}, \tag{4}$$

provided only that x is relatively prime to m.

To prove this, it is only necessary to modify Ivory's method by omitting from the numbers 0, 1, ... , $m-1$ not only the number 0, but all numbers which are not relatively prime to m. There remain $\phi(m)$ numbers, say

$$a_1, a_2, \ldots, a_\mu, \quad \text{where } \mu = \phi(m).$$

Then the numbers

$$a_1 x, a_2 x, \ldots, a_\mu x$$

are congruent, in some order, to the previous numbers, and on multiplying and cancelling a_1, a_2, ... , a_μ (as is permissible) we obtain $x^\mu \equiv 1 \pmod{m}$, which is (4).

To illustrate this proof, take $m = 20$. The numbers less than 20 and relatively prime to 20 are

$$1, 3, 7, 9, 11, 13, 17, 19,$$

so that $\phi(20) = 8$. If we multiply these by any number x which is relatively prime to 20, the new numbers are congruent to the original numbers in some other order. For example, if x is 3, the new numbers are congruent respectively to

$$3, 9, 1, 7, 13, 19, 11, 17 \text{ (mod 20)};$$

and the argument proves that $3^8 \equiv 1$ (mod 20). In fact, $3^8 = 6561$.

4. Euler's function $\phi(m)$

As we have just seen, this is the number of numbers up to m that are relatively prime to m. It is natural to ask what relation $\phi(m)$ bears to m. We saw that $\phi(p) = p - 1$ for any prime p. It is also easy to evaluate $\phi(p^a)$ for any prime power p^a. The only numbers in the set $0, 1, 2, \dots, p^a - 1$ which are not relatively prime to p are those that are divisible by p. These are the numbers pt, where $t = 0, 1, \dots,$ $p^{a-1} - 1$. The number of them is p^{a-1}, and when we subtract this from the total number p^a, we obtain

$$\phi(p^a) = p^a - p^{a-1} = p^{a-1}(p-1). \tag{5}$$

The determination of $\phi(m)$ for general values of m is effected by proving that this function is *multiplicative*. By this is meant that if a and b are any two *relatively prime* numbers, then

$$\phi(ab) = \phi(a)\phi(b). \tag{6}$$

To prove this, we begin by observing a general principle: *if a and b are relatively prime, then two simultaneous congruences of the form*

$$x \equiv \alpha \text{ (mod } a), \quad x \equiv \beta \text{ (mod } b) \tag{7}$$

are precisely equivalent to one congruence to the modulus ab. For the first congruence means that $x = \alpha + at$ where t is an integer. This satisfies the second congruence if and only if

$$\alpha + at \equiv \beta \text{ (mod } b), \quad \text{or} \quad at \equiv \beta - \alpha \text{ (mod } b).$$

This, being a linear congruence for t, is soluble. Hence the two congruences (7) are simultaneously soluble. If x and x' are two solutions, we have $x \equiv x'$ (mod a) and $x \equiv x'$ (mod b), and therefore $x \equiv x'$ (mod ab). Thus there is exactly one solution to the modulus ab. This principle, which extends at once to several congruences, provided that the moduli are relatively prime in pairs, is sometimes called 'the Chinese remainder theorem'. It assures us of the existence of numbers which leave prescribed remainders on division by the moduli in question.

Let us represent the solution of the two congruences (7) by

$$x \equiv [\alpha, \beta] \quad (\text{mod } ab),$$

so that $[\alpha, \beta]$ is a certain number depending on α and β (and also on a and b of course) which is uniquely determined to the modulus ab. Different pairs of values of α and β give rise to different values for $[\alpha, \beta]$. If we give α the values $0, 1, \ldots, a-1$ (forming a complete set of residues to the modulus a) and similarly give β the values $0, 1, \ldots, b-1$, the resulting values of $[\alpha, \beta]$ constitute a complete set of residues to the modulus ab.

It is obvious that if α has a factor in common with a, then x in (7) will also have that factor in common with a, in other words, $[\alpha, \beta]$ will have that factor in common with a. Thus $[\alpha, \beta]$ will only be relatively prime to ab if α is relatively prime to a and β is relatively prime to b, and conversely these conditions will ensure that $[\alpha, \beta]$ is relatively prime to ab. It follows that if we give α the $\phi(a)$ possible values that are less than a and prime to a, and give β the $\phi(b)$ values that are less than b and prime to b, there result $\phi(a)\phi(b)$ values of $[\alpha, \beta]$, and these comprise all the numbers that are less than ab and relatively prime to ab. Hence $\phi(ab)=\phi(a)\phi(b)$, as asserted in (6).

To illustrate the situation arising in the above proof, we tabulate below the values of $[\alpha, \beta]$ when $a=5$ and $b=8$. The possible values for α are $0, 1, 2, 3, 4$, and the possible values for β are $0, 1, 2, 3, 4, 5, 6, 7$. Of these there are four values of α which are relatively prime to a, corresponding to the fact that $\phi(5)=4$, and four values of β that are relatively prime to b, corresponding to the fact that $\phi(8)=4$, in accordance with the formula (5). These values are italicized, as also are the corresponding values of $[\alpha, \beta]$. The latter constitute the sixteen numbers that are relatively prime to 40 and less than 40, thus verifying that $\phi(40)=\phi(5)\phi(8)=4\times4=16$.

α \ β	*0*	*1*	*2*	*3*	*4*	*5*	*6*	*7*
0	0	25	10	35	20	5	30	15
1	16	*1*	26	*11*	36	*21*	6	*31*
2	32	17	2	27	12	37	22	7
3	8	*33*	18	*3*	28	*13*	38	*23*
4	24	9	34	19	4	29	14	39

We now return to the original question, that of evaluating $\varphi(m)$ for any number m. Suppose the factorization of m into prime powers is

$$m = p^a q^b \dots .$$

Then it follows from (5) and (6) that

$$\phi(m) = (p^a - p^{a-1})(q^b - q^{b-1}) \dots ,$$

or, more elegantly,

$$\phi(m) = m(1 - \tfrac{1}{p})(1 - \tfrac{1}{q}) \dots . \tag{8}$$

For example,

$$\phi(40) = 40(1 - \tfrac{1}{2})(1 - \tfrac{1}{5}) = 16,$$

and

$$\phi(60) = 60(1 - \tfrac{1}{2})(1 - \tfrac{1}{3})(1 - \tfrac{1}{5}) = 16.$$

The function $\phi(m)$ has a remarkable property, first given by Gauss in his *Disquisitiones*. It is that the sum of the numbers $\phi(d)$, extended over all the divisors d of a number m, is equal to m itself. For example, if $m = 12$, the divisors are 1, 2, 3, 4, 6, 12, and we have

$$\phi(1) + \phi(2) + \phi(3) + \phi(4) + \phi(6) + \phi(12)$$
$$= 1 + 1 + 2 + 2 + 2 + 4 = 12.$$

A general proof can be based either on (8), or directly on the definition of the function.

We have already referred (I.5) to a table of the values of $\phi(m)$ for $m \leq 10,000$. The same volume contains a table giving those numbers m for which $\phi(m)$ assumes a given value up to 2,500. This table shows that, up to that point at least, every value assumed by $\phi(m)$ is assumed at least twice. It seems reasonable to conjecture that this is true generally, in other words that *for any number m there is another number m' such that* $\phi(m') = \phi(m)$. This has never been proved, and any attempt at a general proof seems to meet with formidable difficulties. For some special types of numbers the result is easy, e.g. if m is odd, then $\phi(m) = \phi(2m)$; or again if m is not divisible by 2 or 3 we have $\phi(3m) = \phi(4m) = \phi(6m)$.

5. Wilson's theorem

This theorem was first published by Waring in his *Meditationes Algebraicae* of 1770, and was ascribed by him to Sir John Wilson

(1741–93), a lawyer who had studied mathematics at Cambridge. It asserts that

$$(p-1)! \equiv -1 \quad (\text{mod } p) \qquad (9)$$

for any prime p.

The following simple proof was given by Gauss. It is based on associating each of the numbers $1, 2, \ldots, p-1$ with its reciprocal (mod p), in the sense defined in §2. The reciprocal of a means the number a' for which $aa' \equiv 1$ (mod p). Each number in the set $1, 2, \ldots, p-1$ has exactly one reciprocal in the set. The reciprocal of a may be the same as a itself, but this only happens if $a^2 \equiv 1$ (mod p), that is, if $a \equiv \pm 1$ (mod p), which requires $a=1$ or $p-1$. Apart from these two numbers, the remaining numbers $2, 3, \ldots, p-2$ can be paired off so that the product of those in any pair is $\equiv 1$ (mod p). It follows that

$$2 \times 3 \times 4 \times \ldots \times (p-2) \equiv 1 \,(\text{mod } p).$$

Multiplying by $p-1$, which is $\equiv -1$ (mod p), we obtain the result (9). The proof just given fails if p is 2 or 3, but it is immediately verified that the result is still true.

Wilson's theorem is one of a series of theorems which relate to the symmetrical functions of the numbers $1, 2, \ldots, p-1$. It asserts that the product of these numbers is congruent to -1 (mod p). Many results are also known concerning other symmetrical functions. As an illustration, consider the sum of the kth powers of these numbers:

$$S_k = 1^k + 2^k + \ldots + (p-1)^k,$$

where p is a prime greater than 2. It can be proved that $S_k \equiv 0$ (mod p) except when k is a multiple of $p-1$. In the latter case, each term in the sum is $\equiv 1$ by Fermat's theorem, and there are $p-1$ terms, so that the sum is $\equiv p-1 \equiv -1$ (mod p).

6. Algebraic congruences

The analogy between congruences and equations suggests the consideration of algebraic congruences, that is, congruences of the form

$$a_n x^n + a_{n-1} x^{n-1} + \ldots + a x_1 + a_0 \equiv 0 \,(\text{mod } m), \qquad (10)$$

where $a_n, a_{n-1}, \ldots, a_0$ are given integers, and x is an unknown. It is naturally an interesting question how far the theory of algebraic equations applies to algebraic congruences, and in fact the study of algebraic congruences constitutes (in various forms) an important part of the theory of numbers.

If n, the degree of the congruence, is 1, (10) reduces to $a_1 x + a_0 \equiv 0 \pmod{m}$, which is a linear congruence of the kind considered in §2.

If a number x_0 satisfies an algebraic congruence to the modulus m, then so does any number x which is congruent to $x_0 \pmod{m}$. Hence two congruent solutions can be considered as the same, and in counting the *number* of solutions of a congruence, we count the number in some complete set of residues (mod m), for example in the set $0, 1, \ldots, m-1$. The congruence $x^3 \equiv 8 \pmod{13}$ is satisfied when $x \equiv 2$ or 5 or 6 (mod 13), and not otherwise, and therefore has three solutions.

We begin by establishing an important principle concerning algebraic congruences. This is that in order to determine the number of solutions of such a congruence, it suffices to treat the case when the modulus is a power of a prime.

To see why this is so, let us suppose that the modulus m can be factorized as $m_1 m_2$, where m_1 and m_2 are relatively prime. An algebraic congruence

$$f(x) \equiv 0 \pmod{m} \qquad (11)$$

is satisfied by a number x if and only if both the congruences

$$f(x) \equiv 0 \pmod{m_1} \quad \text{and} \quad f(x) \equiv 0 \pmod{m_2} \qquad (12)$$

are satisfied. If either of these is insoluble, then the given congruence is insoluble. If both these are soluble, denote the solutions of the former by

$$x \equiv \xi_1, \, x \equiv \xi_2, \, \ldots \pmod{m_1}$$

and those of the latter by

$$x \equiv \eta_1, \, x \equiv \eta_2, \, \ldots \pmod{m_2}.$$

Each solution of (11) corresponds to some one of the ξ's and some one of the η's. Conversely, if we select one of the ξ's, say ξ_i, and one of the η's, say η_j, the simultaneous congruences

$$x \equiv \xi_i \pmod{m_1} \quad \text{and} \quad x \equiv \eta_j \pmod{m_2}$$

are equivalent, as we saw in the last section, to exactly one congruence to the modulus m. It follows that if $N(m)$ denotes the number of solutions of the congruence (11), and $N(m_1)$ and $N(m_2)$ denote the numbers of solutions of the two congruences (12), then

$$N(m) = N(m_1)N(m_2).$$

In other words, $N(m)$ is a multiplicative function of m. If m is factorized into prime powers in the usual form, then

$$N(m) = N(p^a)N(q^b) \dots . \tag{13}$$

That is, if we know the number of solutions of an algebraic congruence for every prime power modulus we can deduce the number of solutions for a general modulus by multiplication. In particular, if one of the numbers $N(p^a)$ is zero for one of the prime powers composing m, then the congruence is insoluble, as is of course obvious.

A similar result holds for algebraic congruences in more than one unknown. The number of solutions of a congruence

$$f(x, y) \equiv 0 \pmod{m}$$

in two unknowns (and similarly in any number of unknowns) is again a multiplicative function of the modulus.

7. Congruences to a prime modulus

There are two reasons why the theory of congruences is largely concerned with congruences to prime moduli. As we have just seen, it suffices in determining the number of solutions of a congruence to consider the case when the modulus is a prime power. It so happens that the behaviour of a congruence to a prime power modulus p^a is often deducible from its behaviour in the case when the modulus is simply p. Consequently a theory of congruences to a prime modulus is the first essential.

The second reason lies in the specially simple nature of arithmetic to a prime modulus, which was already pointed out in §2. In this arithmetic we have p elements, represented by the numbers $0, 1, 2, \dots, p-1$, which can be combined by all the four operations of addition, multiplication, subtraction and division, apart from division by zero. The first three operations are carried out as usual, except that the resulting number is brought back into the set by adding or subtracting the appropriate multiple of p; the last operation, that of division, is carried out by solving a linear congruence.

A set of elements (of what nature is immaterial) which can be combined by operations analogous to the four operations of arithmetic and satisfying the same laws, and such that all four operations can always be carried out within the system, except for the operation of division by the zero element, is called a *field*. The most familiar example of a field is provided by the system of rational numbers. But the numbers $0, 1, \ldots, p-1$, when combined as explained above, also form a field, and though this is a less familiar example it is simpler in that the field comprises only a finite number of elements. The simplest case of all occurs when $p=2$. We then have an arithmetic with two elements. If we call them O and I (corresponding to 0 and 1), the rules of calculation are:

$$O+O=O, \ O+I=I, \ I+O=I, \ I+I=O;$$
$$O\times O=O, \ O\times I=O, \ I\times O=O, \ I\times I=I.$$

One way of describing this arithmetic is to say that it is the degenerate form of ordinary arithmetic in which every even number has been replaced by O and every odd number by I.

There are some theorems of elementary algebra which are valid when the symbols represent elements of any field. One of these is the theorem that an algebraic equation of degree n has at most n solutions. In particular, this theorem is valid in the mod p field, where it takes the form that *a congruence of degree n, say*

$$a_n x^n + a_{n-1} x^{n-1} + \ldots + a_1 x + a_0 \equiv 0 \ (\text{mod } p), \tag{14}$$

cannot have more than n solutions. It is to be understood that the highest coefficient a_n is not congruent to 0 (mod p) since if it were the term would be omitted.

This result was first stated and proved by Lagrange in 1768. The proof is the same as that of the corresponding result for equations. The essential point is that if x_1 is any solution of the congruence, the polynomial on the left-hand side of the congruence factorizes, one of the factors being the linear polynomial $x-x_1$. For if x_1 satisfies the congruence, we have

$$a_n x_1{}^n + a_{n-1} x_1{}^{n-1} + \ldots + a_1 x_1 + a_0 \equiv 0 \ (\text{mod } p).$$

If we subtract this from (14), each difference of corresponding terms is of the form $a_k(x^k - x_1{}^k)$, where k is one of the numbers $0, 1, \ldots, n$. Each such difference contains the linear polynomial $x-x_1$ as a factor. Thus the congruence (14) can be written in the form

$$(x-x_1)(b_{n-1}x^{n-1}+b_{n-2}x^{n-2}+ \ldots +b_0)\equiv 0 \text{ (mod } p),$$

where $b_{n-1}, \ldots , b_0$ are certain integers depending on $a_n, \ldots a_0$ and on x_1. Any other solution, say x_2, of the congruence (14) must (since p is a prime) satisfy

$$b_{n-1}x^{n-1}+b_{n-2}x^{n-2}+ \ldots +b_0\equiv 0 \text{ (mod } p),$$

and must give rise to a factor $x-x_2$ of the polynomial here, so that we then have two linear factors for the original polynomial. This goes on until either the left-hand side of (14) is completely factorized, or we come to a congruence which is insoluble. In the former case, the congruence (14) has exactly n solutions, in the latter case it has fewer than n solutions.

It is essential for Lagrange's theorem that the modulus should be a prime. For example, the congruence $x^2-1\equiv 0$ (mod 8), though of degree 2, has the four solutions $x\equiv 1, 3, 5, 7$ (mod 8), being in fact satisfied by every odd number.

We have seen that each solution of an algebraic congruence corresponds to a linear factor of the polynomial in the congruence. One can consider more generally the question of factorizing a polynomial, whose coefficients are integers taken to the modulus p, into other polynomials. It is readily seen that any polynomial $f(x)$ can be factorized into *irreducible* polynomials, that is, poly- nomials which cannot be further factorized. In other words, there exist irreducible polynomials $f_1(x), f_2(x), \ldots , f_r(x)$ such that

$$f(x)\equiv f_1(x)f_2(x) \ldots f_r(x) \text{ (mod } p)$$

identically in x. It will, of course, be appreciated that the irreduci- bility in question here is one which is *relative* to the prime p. Any linear polynomials that may occur in the factorization will corres- pond to solutions of the congruence $f(x)\equiv 0$ (mod p), and if there are no linear factors the congruence is insoluble. Two examples of factorization into irreducible polynomials are:

$$x^4+3x^2+3\equiv (x-1)(x+1)(x^2-3) \quad \text{(mod 7)},$$
$$x^4+2x^3-x^2-2x+2\equiv (x^2+x+1)(x^2+x+2) \quad \text{(mod 5)}.$$

The question arises whether such a factorization is unique. There is the obvious possibility of introducing numerical factors into the polynomials $f_1(x), \ldots , f_r(x)$; provided their product is $\equiv 1$ (mod p) they will have no effect. It can be proved that apart from this possibility, *the factorization is unique.* The theory is very similar to that of the factorization of the natural numbers into primes. An important part is again played by Euclid's algorithm, which is now based on the process for dividing one polynomial by another with a remainder whose degree is less than the degree of the divisor. Lack of space precludes us from giving any further account of this theory.

8. Congruences in several unknowns

A very simple and general theorem, due to Chevalley, establishes the solubility of a wide class of congruences in several unknowns. Suppose $f(x_1, x_2, \dots, x_n)$ is any polynomial in n variables, not necessarily homogeneous, whose degree is less than n, and in which the constant term is zero. By the *degree* is to be understood the highest degree of any individual term, where the degree of a term such as $x_1 x_2{}^3 x_3{}^4$ is taken to be $1+3+4=8$. Chevalley's theorem is that the congruence

$$f(x_1, x_2, \dots, x_n) \equiv 0 \pmod{p} \tag{15}$$

is necessarily soluble, with not all the unknowns congruent to zero.

Before giving the proof, there is one preliminary remark which is relevant. Under what circumstances can a congruence, say

$$\varphi(x_1, x_2, \dots, x_n) \equiv 0 \pmod{p},$$

hold for *all* integers $x_1, x_2, \dots, x_n$? By Fermat's theorem (§3) we have $x^p \equiv x \pmod{p}$ for all x. Therefore in any congruence each exponent in each term can be reduced to one of the values $1, 2, \dots, p-1$, by subtracting a multiple of $p-1$, without affecting the significance of the congruence. After this has been done, the resulting congruence can only hold for all integers $x_1, x_2, \dots, x_n$ if it reduces to an identity, that is, if all the coefficients in the new congruence are congruent to zero. For Lagrange's theorem tells us that such a congruence, of degree at most $p-1$ in x_1, can have at most $p-1$ solutions for x_1, unless all its coefficients (when it is regarded as a polynomial in x_1) are congruent to zero. These coefficients are polynomials in $x_2, \dots, x_n$ of degree at most $p-1$ in each unknown, and we can apply the same argument to these polynomials. The general proposition follows, on repetition of the argument.

Chevalley's theorem is proved by deriving from the congruence (15), which is supposed not to be satisfied except when the unknowns are all zero, another congruence which is satisfied for *all* values of the unknowns. This is the congruence

$$1 - [f(x_1, \dots, x_n)]^{p-1} \equiv (1 - x_1^{p-1}) \dots (1 - x_n^{p-1}) \pmod{p}. \quad (16)$$

If $x_1, \dots, x_n$ are all congruent to zero, both sides are congruent to 1. If any one of $x_1, \dots, x_n$ is not congruent to zero, the left-hand side is congruent to zero by Fermat's theorem, and so also is the right-hand side. Hence, on the hypothesis which is to be disproved, (16) holds for all integers $x_1, \dots, x_n$. By what we have seen above, the relation must reduce to an identity if, after writing out all the terms, we reduce each exponent of each variable to one of the values 1, 2, ... , $p-1$ by subtracting a suitable multiple of $p-1$. On the right, no such reduction is possible, since each individual exponent is already at most $p-1$. On the left, reduction may be possible. But the total degree of each term on the left is less than $(p-1)n$ by hypothesis, and reduction of exponents can only diminish this degree. It now becomes plain that the relation cannot reduce to an identity, since no term on the left will be of as high a degree as the term $x_1^{p-1} x_2^{p-1} \dots x_n^{p-1}$ on the right. This proves the theorem. As a simple illustration, we may take the congruence

$$x^2 + y^2 + z^2 \equiv 0 \pmod{p}.$$

The left-hand side is of degree 2 in the 3 variables x, y, z, and has no constant term, so the hypotheses are satisfied. It follows that the congruence is soluble, with x, y, z not all congruent to zero. This particular result is useful in connection with the representation of a number as a sum of four squares (V.4), though when needed for that purpose it can also be easily proved directly.

9. *Congruences covering all numbers*

A curious problem is that of finding sets of congruences, to distinct moduli, such that every number satisfies one at least of the congruences. Such a set of congruences may be called a covering set. Naturally the modulus 1 must be excluded. The congruences

$$x \equiv 0 \pmod{2}, \ 0 \pmod{3}, \ 1 \pmod{4}, \ 1 \pmod{6}, \ 11 \pmod{12}$$

constitute a covering set. For the first two cover all numbers except those congruent to 1 or 5 or 7 or 11 (mod 12). Of these, 1 and 5 are covered by $x \equiv 1 \pmod 4$, 7 is covered by $x \equiv 1 \pmod 6$, and 11 is covered by the last congruence.

Erdös has proposed the problem: given any number N, does

there exist a set of covering congruences which uses only moduli greater than N? Probably this is true, but it is not easy to see how to give a proof. Erdös himself has given a set which does not use the modulus 2, the moduli being various factors of 120. Church-house has given a set for which the least modulus is 9; here the moduli are various factors of 604,800. Choi has shown that there is a set with least modulus 20. The question whether or not there is a set with every modulus odd is still open.

NOTES

§3. The usual phrase is that 'x belongs to the exponent l with respect to the modulus m', but this seems unnecessarily cumbrous.

§4. The number $[\alpha, \beta]$, introduced to represent the solution of the simultaneous congruences (7), can be expressed by a formula as follows. Determine a' and b' so that $aa' \equiv 1 \pmod{b}$ and $bb' \equiv 1 \pmod{a}$; then $[\alpha, \beta] \equiv aa'\beta + bb'\alpha \pmod{ab}$.

§5. Wilson's theorem can be generalized to the case of a composite modulus; see Hardy and Wright, §8.8, or Ore, p. 266.
The usual proof that $S_k \equiv 0 \pmod{p}$ employs a primitive root, as in Hardy and Wright, §7.10, but more direct proofs can also be given. For the extensive literature on the symmetrical functions of the numbers $1, 2, \ldots, p-1$, see Dickson's *History*, vol. 1, ch. 3.

§7. The complete determination of all types of field consisting of a finite number of elements was made by the American mathematician E. H. Moore in 1893. The number of elements is necessarily a prime power p^n, and the field is either the mod p field (when $n=1$) or is an algebraic extension of it. For accounts of the theory, see Dickson, *Linear Groups* (Teubner), ch. 1, or MacDuffee, *Introduction to Abstract Algebra* (Wiley), pp. 174–80, or Birkhoff and MacLane, *Survey of Modern Algebra* (Macmillan, New York), pp. 428–31. For some tables of irreducible polynomials for the first four prime moduli, see R. Church, *Annals of Math.*, 36 (1935), 198–209.

§8. For Chevalley's theorem, see *Abhandlungen Math. Seminar Hamburg* 11 (1936), 73–5. Chevalley proved more generally that several simultaneous congruences, which are satisfied when all the variables are 0, will have another solution provided the sum of their degrees is less than the number of variables. In the paper which follows Chevalley's, E. Warning proved that under the same conditions the total number of solutions is divisible by p.

§9. For further work on the subject of covering congruences, see Guy, section F.13. For Choi's construction, see *Math. Comput.*, 25 (1971), 885–95.

III

QUADRATIC RESIDUES

1. *Primitive roots*

In this chapter we shall investigate algebraic congruences to a prime modulus, which contain two terms only, that is, one term besides the constant term. Such a *binomial* congruence can be written in the form

$$ax^k \equiv b \pmod{p}$$

where k, the degree of the congruence, is a positive integer. If a' denotes the reciprocal of a to the modulus p, so that $aa' \equiv 1 \pmod{p}$, and we multiply the above congruence throughout by a', we obtain

$$x^k \equiv a'b \pmod{p}.$$

We can therefore reduce any binomial congruence to one of the simpler form

$$x^k \equiv c \pmod{p}. \tag{1}$$

A number c for which the congruence (1) is soluble is called a kth *power residue to the modulus p*, and similarly if the congruence is insoluble c is said to be a kth *power non-residue*. (It is convenient, however, not to classify numbers c that are congruent to 0 $\pmod{p}$ as kth power residues, even though the congruence is then soluble.) If k is 2 we have quadratic residues and non-residues, and as the theory can be carried further in this case than in the general case we shall later in the chapter consider mainly this possibility.

To illustrate the definition, take p to be 13 and k to be 2 or 3. The values of x^2 and x^3 to the modulus 13 are given below:

x:	1	2	3	4	5	6	7	8	9	10	11	12
x^2:	1	4	9	3	12	10	10	12	3	9	4	1
x^3:	1	8	1	12	8	8	5	5	1	12	5	12.

Thus, to the modulus 13, the numbers 1, 3, 4, 9, 10, 12 are quadratic residues and the remaining numbers, 2, 5, 6, 7, 8, 11, are quadratic non-residues. The numbers 1, 5, 8, 12 are cubic residues, and the remaining numbers, 2, 3, 4, 6, 7, 9, 10, 11, are cubic non-residues.

The theory of kth power residues and non-residues is bound up with the concept of the *order* of a number to the modulus p, which was defined in II.3. The order of any number a, supposed not to be congruent to 0, is the least natural number l for which $a^l \equiv 1$ (mod p). We proved that l is always a factor of $p-1$, and in an example with $p=11$ we found that the order of the number 2 was actually equal to $p-1$. Euler was the first to state that *for any prime p there is some number whose order is equal to $p-1$*, and he called such a number a *primitive root* for the prime p. But his proof of the existence of a primitive root was defective, and the first satisfactory proof was that of Legendre. This proof we now proceed to give.

The first step in the proof is to establish a general principle concerning the order of the product of two numbers. If a number a has the order l, and a number b has the order k, then the number ab has the order lk, *provided* that l and k are relatively prime. Certainly the number ab, when raised to the power lk, gives 1 (mod p), because

$$(ab)^{lk} \equiv (a^l)^k (b^k)^l \equiv 1 \quad (\text{mod } p),$$

since $a^l \equiv 1$ and $b^k \equiv 1$. This fact does not depend on l and k being relatively prime, but it shows only that the order of ab is a divisor of lk. There is still the possibility that it might be a proper divisor of lk, and this we have to exclude. Suppose the order of ab is $l_1 k_1$, where l_1 is a divisor of l and k_1 is a divisor of k. Then

$$a^{l_1 k_1} b^{l_1 k_1} \equiv 1 \ (\text{mod } p).$$

Raise both sides of this congruence to the power l_2, where $l_1 l_2 = l$. Since $a^l \equiv 1$, we obtain $b^{lk_1} \equiv 1$. This implies that lk_1 is a multiple of the order of b, which is k. Since l is relatively prime to k it follows that k_1 is a multiple of k, and being also a divisor of k it must equal k. Similarly $l_1 = l$, and so the order of ab is exactly lk.

The above principle allows one to construct a primitive root step by step. Let $p-1$ be factorized into prime powers, say as

$$p-1 = q_1{}^{a_1} q_2{}^{a_2} \dots . \tag{2}$$

If we can find a number x_1 whose order is $q_1{}^{a_1}$, and a number x_2 whose order is $q_2{}^{a_2}$, and so on, then by repeated application of the principle the product of all these numbers will have the order $p-1$, and will be a primitive root. Hence it remains only to prove that *if q^a is one of the prime powers composing $p-1$, then there is some number whose order (mod p) is exactly q^a.*

A number whose order is q^a must satisfy the congruence

$$x^{q^a} \equiv 1 \pmod{p}. \tag{3}$$

But a number which satisfies this congruence need not have the order q^a; its order may be any factor of q^a, that is, it may be 1 or q or q^2, and so on up to q^{a-1}. However, if the order is not q^a, it will be a factor of q^{a-1}, and the number will satisfy the congruence

$$x^{q^{a-1}} \equiv 1 \pmod{p}. \tag{4}$$

Therefore we need a number which satisfies the congruence (3) but does not satisfy the congruence (4).

We can prove that there is such a number by finding out how many solutions these congruences have. Certainly, by Lagrange's theorem, the congruence (3) has at most q^a solutions, and the congruence (4) has at most q^{a-1} solutions. This in itself would not help us, but fortunately we can prove that these congruences have *exactly* q^a and q^{a-1} solutions. It will follow that there are $q^a - q^{a-1}$ numbers which satisfy (3) and not (4), and since $q^a > q^{a-1}$ this will give what we want, and will complete the proof.

We consider, more generally, the congruence

$$x^d - 1 \equiv 0 \pmod{p},$$

where d is any factor of $p-1$. By Lagrange's theorem, this congruence has at most d solutions, and we shall prove that it has exactly d solutions. The proof depends on the fact that the polynomial $x^d - 1$ is a factor of the polynomial $x^{p-1} - 1$. If we write, for the moment, y in place of x^d, and put $p-1 = de$, then

$$x^{p-1} - 1 = y^e - 1 = (y-1)(y^{e-1} + y^{e-2} + \dots + 1).$$

Since $y-1=x^d-1$, this gives an identity of the form

$$x^{p-1}-1=(x^d-1)f(x),$$

where $f(x)$ is a certain polynomial in x of degree $p-1-d$. Now the congruence

$$x^{p-1}-1\equiv 0 \pmod{p}$$

has $p-1$ solutions, being satisfied by all x not congruent to 0 (II.3). All the $p-1$ solutions must satisfy

$$\textit{either}\quad x^d-1\equiv 0 \pmod{p}\quad \textit{or}\quad f(x)\equiv 0 \pmod{p}.$$

The latter of these has at most $p-1-d$ solutions, by Lagrange's theorem, hence the former must have *at least* d solutions, and therefore has *exactly* d solutions. Taking d to be q^a or q^{a-1}, we obtain what was required in the previous proof.

We illustrate the proof by taking $p=19$. Here $p-1=2\times 3^2$. We require first a number x_1 of order 2, that is a number which satisfies $x^2\equiv 1$, $x\not\equiv 1$. Obviously x_1 must be -1, or (what is the same) 18. We require secondly a number x_2 of order 9, that is a number which satisfies $x^9\equiv 1$ and $x^3\not\equiv 1$. It will be found that the solutions of $x^9\equiv 1 \pmod{19}$ are 1, 4, 5, 6, 7, 9, 11, 16, 17. Of these, the numbers 1, 7, 11 must be ruled out because they satisfy $x^3\equiv 1$. This leaves six choices for x_2, corresponding to q^a-q^{a-1} choices in the general case. Multiplying by x_1 we obtain the primitive roots $-4, -5, -6, -9, -16, -17$, or, what is the same, 2, 3, 10, 13, 14, 15. To verify that 2 is a primitive root, we note that the successive powers of 2 to the modulus 19 are 2, 4, 8, 16, 13, 7, 14, 9, 18, 17, 15, 11, 3, 6, 12, 5, 10, 1, and the first of these which is 1 is the eighteenth. The above method is not a very practical one for finding a primitive root; it is much easier to proceed by trying the numbers 2, 3, ... in succession. But that, of course, would not lead to any general proof of the existence of a primitive root.

It will be seen that the construction in the general proof gives possibly

$$(q_1{}^{a_1}-q_1{}^{a_1-1})(q_2{}^{a_2}-q_2{}^{a_2-1}) \dots$$

primitive roots, by multiplying together all possible values for $x_1, x_2, \dots$. The primitive roots found in this way are in fact all different, and constitute all the primitive roots, but we shall not stop to prove this. The number of primitive roots is given by the above product, whose value is $\phi(p-1)$, by (8) of Chapter II. When $p=19$, for instance, there are $\phi(18)=6$ primitive roots.

2. Indices

The existence of a primitive root is not only of theoretical interest, but provides one with a new tool for use in calculations to a prime modulus p. This tool is very similar to that provided by logarithms in ordinary arithmetic.

Let g be a primitive root mod p. Then the numbers

$$g, g^2, g^3, \dots, g^{p-1} \;(\equiv 1) \tag{5}$$

are all incongruent, since g^{p-1} is the *first* power of g which is congruent to 1. Also none of these numbers is $\equiv 0$. Hence they must be congruent to the numbers $1, 2, \dots, p-1$ in some order. The example in the last section illustrates this; the powers of 2 from 2 itself up to 2^{18} ($\equiv 1$) are congruent to $1, 2, \dots, 18$ to the modulus 19, in another order.

Any number not congruent to 0 (mod p) is therefore congruent to one of the numbers in the series (5). If $a \equiv g^\alpha$ (mod p), we say that α is the *index* of a (relative to the primitive root g). When a is given, this defines α uniquely as one of the numbers $1, 2, \dots, p-1$. But there is no need to restrict α to these values. If α' is any other number for which $a \equiv g^{\alpha'}$, we can reduce α' to one of the set just mentioned by adding or subtracting a multiple of $p-1$, and this does not alter $g^{\alpha'}$ since $g^{p-1} \equiv 1$. The reduced value of α' must be α, and therefore

$$\alpha' \equiv \alpha \;(\text{mod } p-1).$$

If $p=19$ and $g=2$, the indices of the numbers $1, \dots, 18$ are:

number:	1	2	3	4	5	6	7	8	9
index:	18	1	13	2	16	14	6	3	8

number:	10	11	12	13	14	15	16	17	18
index:	17	12	15	5	7	11	4	10	9

To construct such a table, we place the index 1 under the primitive root itself (2 here), then the index 2 under the square of the primitive root (4 here) and so on, calculating the powers of the primitive root to the modulus p (19 here). A table of indices for all primes less than 1,000 was published by Jacobi in 1839, under the title *Canon Arithmeticus*.

By the use of indices one can reduce the operation of multiplication (mod p) to the operation of addition, just as by the use of logarithms one can reduce ordinary multiplication (provided only positive numbers are involved) to addition. If a and b are two given numbers, and α and β are their indices, then $a \equiv g^\alpha$ and $b \equiv g^\beta$, whence $ab \equiv g^{\alpha+\beta}$, all these congruences being to the modulus p. It follows that the index of the product ab is either equal to $\alpha + \beta$ or differs from it by a multiple of $p-1$. Thus to multiply two numbers together, one looks up their indices in the table, adds them, then brings the result to lie in the range 1, 2, ... , $p-1$ by subtracting a multiple of $p-1$ if necessary; then looks up the number having this index. For example, to find the value of 10×12 (mod 19), we see that the indices of these numbers in the above table are 17 and 15 respectively; the sum is 32, which is equivalent to 14 on subtracting 18 ($=p-1$); the number whose index is 14 is 6, and therefore this is the answer. One can carry out division (mod p) in the same way as multiplication, except that one subtracts the indices instead of adding them.

The use of indices enables us to investigate the structure of the kth power residues and non-residues (mod p). We wish to decide whether the congruence

$$x^k \equiv a \pmod p \tag{6}$$

is soluble or insoluble. If the index of x is ξ, the index of x^k is $k\xi$, or differs from this by a multiple of $p-1$. Hence the above congruence is equivalent to

$$k\xi \equiv \alpha \pmod{p-1}, \tag{7}$$

where α is the index of a. This is a linear congruence for the unknown ξ to the modulus $p-1$.

If k is relatively prime to $p-1$ the position is very simple: the linear congruence (7) has a unique solution for ξ, and the congruence (6) therefore has a unique solution for x. Every number is a kth power residue, and in exactly one way. In other words, if k is relatively prime to $p-1$, the numbers

$$1^k, 2^k, 3^k, \ldots , (p-1)^k$$

are congruent to the numbers 1, 2, ... , $p-1$, in some other order. For example, if p is 19 and k is 5, the numbers $1^5, 2^5, \ldots , 18^5$ are congruent (mod 19) to

$$1, 13, 15, 17, 9, 5, 11, 12, 16, 3, 7, 8, 14, 10, 2, 4, 6, 18.$$

The position is quite different if k has a factor in common with $p-1$. Let us first look at a particular case, say $p=19$ and $k=3$. The congruence (7) is now

$$3\xi \equiv \alpha \;(\text{mod } 18).$$

This congruence is obviously insoluble unless α is divisible by 3. If α is divisible by 3, say $\alpha = 3\beta$, the last congruence becomes $\xi \equiv \beta \;(\text{mod } 6)$. This gives one value for ξ to the modulus 6, but *three* values to the modulus 18, which is the appropriate modulus for ξ, namely β, $\beta + 6$, $\beta + 12$ if β is one solution. Thus, if α is divisible by 3, the number a is congruent to three distinct cubes. Looking at the table of indices to the modulus 19, we see that the numbers whose indices are divisible by 3 are 1, 7, 8, 11, 12, 18. If a is one of these numbers, the congruence $x^3 \equiv a \;(\text{mod } p)$ has exactly three solutions. These numbers are the cubic residues $(\text{mod } 19)$, and the remaining 12 numbers are cubic non-residues.

The general situation can be investigated in the same way. Let K denote the highest common factor of k and $p-1$. The congruence (7) is insoluble for ξ if α is not a multiple of K, since k and the modulus are both divisible by K. On the other hand, if α is a multiple of K the congruence (7) is soluble for ξ, and has exactly K solutions. Thus *the kth power residues $(\text{mod } p)$ consist of just those numbers whose indices are divisible by K, the highest common factor of k and $p-1$.* If a is a kth power residue, the congruence (6) has exactly K solutions. The number of kth power residues is $\dfrac{p-1}{K}$, since the possible indices are the numbers $1, 2, \ldots, p-1$, and a proportion $\dfrac{1}{K}$ of these numbers are divisible by K.

The simplest case is $k = 2$, when we are concerned with quadratic residues and non-residues. If we suppose that $p > 2$ then $p-1$ is even, and the highest common factor of 2 and $p-1$ is itself 2. The conclusion in this case is that *the quadratic residues are the numbers with even indices and the quadratic non-residues are the numbers with odd indices. There are equal numbers of them, namely $\frac{1}{2}(p-1)$ of each.* If a is any quadratic residue, the theory tells us that the congruence $x^2 \equiv a \;(\text{mod } p)$ has exactly two solutions. It is plain that if $x \equiv x_1$ is one solution, the other is $x \equiv -x_1$.

If $p = 19$, the quadratic residues are

$$1, 4, 5, 6, 7, 9, 11, 16, 17$$

and the quadratic non-residues are

$$2, 3, 8, 10, 12, 13, 14, 15, 18.$$

3. *Quadratic residues*

For the rest of this chapter, we shall restrict ourselves to the theory of quadratic residues and non-residues, a theory which can be carried considerably further than the general theory of kth power residues. We shall suppose throughout that p is a prime other than 2.

As we have just seen, half the numbers 1, 2, ... , $p-1$ are quadratic residues and the other half are quadratic non-residues. The quadratic residues are congruent to the numbers

$$1^2, 2^2, \ldots , [\tfrac{1}{2}(p-1)]^2;$$

for the remaining numbers, from $\tfrac{1}{2}(p+1)$ to $p-1$, give the same results on squaring, since $(p-x)^2 \equiv x^2 \pmod{p}$.

The quadratic residues and non-residues have a simple multiplicative property; the product of two residues or of two non-residues is a residue, whereas the product of a residue and a non-residue is a non-residue. This follows at once from the fact that the residues have even indices and the non-residues have odd indices: the sum of two even indices or of two odd indices is even, whereas the sum of an even and an odd index is odd. Thus, for example, in the lists of quadratic residues and non-residues for the prime 19, at the end of §2, the product of any two numbers taken from the same list is congruent to a number in the first list, and the product of any two numbers taken from different lists is congruent to one in the second list.

It was doubtless this multiplicative property which suggested to Legendre the introduction of a symbol by which to express the quadratic character of a number a with respect to a prime p. Legendre's symbol is defined as follows:

$$\left(\frac{a}{p}\right) = \begin{cases} 1 & \text{if } a \text{ is a quadratic residue (mod } p), \\ -1 & \text{if } a \text{ is a quadratic non-residue (mod } p). \end{cases}$$

For convenience of printing we shall also use the form $(a|p)$. Another way of expressing the definition is that $(a|p)=(-1)^\alpha$, where α is the index of a. The multiplicative property takes the form

$$\left(\frac{ab}{p}\right) = \left(\frac{a}{p}\right)\left(\frac{b}{p}\right).$$

Every number a (not congruent to 0) satisfies Fermat's congruence $a^{p-1}-1\equiv 0 \pmod{p}$. Since $p-1$ is even, this congruence factorizes, and if we put $p-1=2P$ we can say that every number satisfies

$$\text{either} \quad a^P \equiv 1 \quad \text{or} \quad a^P \equiv -1 \text{ (mod } p\text{)}.$$

Euler was apparently the first to prove that the distinction between these two possibilities corresponds exactly to the distinction between a being a quadratic residue or non-residue. From our present point of view, the proof is immediate. If α is the index of a, then $a^P \equiv g^{\alpha P}$ (mod p). If α is even, αP is a multiple of $p-1$, and $g^{\alpha P} \equiv 1$. If α is odd, $\alpha P = \frac{1}{2}\alpha(p-1)$ is not a multiple of $p-1$, and $g^{\alpha P}$ cannot be congruent to 1, and so must be congruent to -1. The result is called *Euler's criterion* for the quadratic character of a. In terms of Legendre's symbol, it takes the form

$$\left(\frac{a}{p}\right) \equiv a^P \text{ (mod } p\text{)}, \quad \text{where } P = \frac{1}{2}(p-1). \tag{8}$$

Euler's criterion is not in itself of great use in investigating the properties of quadratic residues and non-residues, but it does give at once the rule for the quadratic character of the number -1. The value of $(-1)^P$ will be 1 or -1 according as P is even or odd, that is, according as p is of the form $4k+1$ or $4k+3$. Hence -1 *is a quadratic residue for primes of the form $4k+1$, and a quadratic non-residue for primes of the form $4k+3$*. This means that for a prime of the form $4k+1$, the lists of quadratic residues and non-residues are both symmetrical, that is, the character of $p-a$ is the same as that of a. For $p-a \equiv -a$, and $(-a|p)=(-1|p)(a|p)=(a|p)$. On the other hand, if p is of the form $4k+3$, the character of $p-a$ is opposite to that of a, as may be seen in the case $p=19$ (at the end of §2).

The fact that the congruence $x^2+1 \equiv 0$ (mod p) is soluble for primes of the form $4k+1$ and insoluble for primes of the form $4k+3$ was known to Fermat. It seems to have been first proved by Euler, after repeated failures, in about 1749, whereas he did not discover his criterion until 1755. Lagrange, in 1773, pointed out that there is a very simple way of giving explicitly the solutions of the congruence when it is soluble. If $p=4k+1$, Wilson's theorem (II.5) states that

$$1 \times 2 \times 3 \times \ldots \times 4k \equiv -1 \text{ (mod } p\text{)}.$$

Now $4k \equiv -1$, $4k-1 \equiv -2$, and so on, down to $2k+1 \equiv -2k$. Substituting these values, we get

$$(1 \times 2 \times 3 \times \ \dots \ \times 2k)^2 \equiv -1 \ (\mathrm{mod} \ p),$$

since the number of negative signs introduced is $2k$, and is even. Hence the solutions of the congruence $x^2 \equiv -1$ (mod p) are $x \equiv \pm(2k)!$, where $p = 4k+1$. For example, if $p = 13$, so that $k = 3$, the solutions are

$$x \equiv \pm 6! \equiv \pm 720 \equiv \pm 5 \ (\mathrm{mod} \ 13).$$

Naturally the construction is not a useful one for numerical work, but it is always interesting to have an explicit construction to supplement an existence proof.

4. Gauss's lemma

The deeper properties of quadratic residues and non-residues, especially those associated with the law of reciprocity (§5), were discovered empirically, and the first proofs were by very complicated and indirect methods. It was not until 1808 (seven years after the publication of his *Disquisitiones*) that Gauss discovered a simple lemma, which provides the key to a simple and elementary proof of the law of reciprocity.

Gauss's lemma gives a rule for the quadratic character of a number a (not congruent to 0) with respect to a prime p. As always, we suppose $p > 2$, and put $P = \frac{1}{2}(p-1)$. The rule is to form the numbers

$$a, 2a, 3a, \ \dots \ , Pa, \tag{9}$$

and reduce each of these to lie between $-\frac{1}{2}p$ and $\frac{1}{2}p$, by subtracting the appropriate multiple of p from each one. Let v be the number of *negative* numbers in the resulting set of numbers. Then $(a|p) = (-1)^v$, that is, *a is a quadratic residue if v is even, and a quadratic non-residue if v is odd*. The proof is quite simple. The rule requires us to express each of the numbers in the set (9) as congruent to one of the numbers $\pm 1, \pm 2, \ \dots \ , \pm P$, as we obviously can. When we do this, no number in the set $1, 2, \ \dots \ , P$ occurs more than once, either with positive or with negative sign. For if the same number occurred twice with the same sign, it would mean that two of the numbers in the set (9) were congruent to one another (mod p), which is not the case. If the same number occurred twice with opposite signs, it would mean that the sum of two numbers in the set (9) was congruent to zero (mod p), which is also not the case. So the resulting set consists of the numbers $\pm 1, \pm 2, \ \dots \ , \pm P$, with a certain *definite sign* prefixed to each of them. Multiplying the two sets, we get

$$(a)(2a)(3a) \ldots (Pa) \equiv (\pm 1)(\pm 2)(\pm 3) \ldots (\pm P) \,(\mathrm{mod}\,p).$$

On cancelling 2, 3, ... , P it follows that

$$a^P \equiv (\pm 1)(\pm 1) \ldots (\pm 1) = (-1)^\nu$$

where ν is the number of negative signs. This proves the result, by Euler's criterion (§3).

To illustrate Gauss's lemma numerically, take $p = 19$ and $a = 5$. Here $P = 9$, and we have to reduce the numbers 5, 10, 15, ... , 45 so that they lie between -9 and 9 inclusive. The resulting numbers are

$$5, -9, -4, 1, 6, -8, -3, 2, 7.$$

As in the general theory, these consist of the numbers from 1 to 9, each with a particular sign. The number of negative signs is 4, and since this is even, 5 is a quadratic residue (mod 19), or symbolically: $(5|19) = 1$.

Gauss's lemma enables one to give a simple rule for the quadratic character of 2. When $a = 2$, the series of numbers in (9) is

$$2, 4, 6, \ldots , 2P,$$

and $2P = p - 1$. We have to determine how many of the numbers in this set, when reduced to lie between $-\tfrac{1}{2}p$ and $\tfrac{1}{2}p$, become negative. Since all the numbers are between 0 and p, those which become negative are those greater than $\tfrac{1}{2}p$. So we have merely to find how many numbers of the form $2x$ satisfy $\tfrac{1}{2}p < 2x < p$; in other words, how many integers x there are which satisfy $\tfrac{1}{4}p < x < \tfrac{1}{2}p$. Put $p = 8k + r$, where r is 1 or 3 or 5 or 7. The condition is

$$2k + \tfrac{1}{4}r < x < 4k + \tfrac{1}{2}r,$$

and we wish to know whether the number of integers x satisfying this condition is even or odd. Now the parity of the number will not be changed if we remove the even numbers $2k$ and $4k$ from the two sides of the inequality. Hence it is sufficient to consider the inequality $\tfrac{1}{4}r < x < \tfrac{1}{2}r$. This inequality has no solution if r is 1, one solution if r is 3 or 5, two solutions if r is 7. Hence 2 is a quadratic residue in the first and last cases, and a non-residue in the two middle cases. So the rule is that 2 *is a quadratic residue for primes of the form* $8k \pm 1$, *and a quadratic non-residue for primes of the form* $8k \pm 3$. This fact was known to Fermat, but was first proved, after great difficulty and in a very complicated way, by Euler and Lagrange.

It will be instructive to work out another rule of a similar kind by Gauss's lemma, as the same method will be used in the next section to prove the law of reciprocity. Let us find for what primes 3 is a residue or non-residue. The numbers 3, 6, 9, ... , $3P$ are all less than $\frac{3}{2}p$, and consequently the only ones which become negative, when reduced to lie between $-\frac{1}{2}p$ and $\frac{1}{2}p$, are those between $\frac{1}{2}p$ and p. We require the number of numbers x for which $\frac{1}{2}p < 3x < p$, that is $\frac{1}{6}p < x < \frac{1}{3}p$. Put $p = 12k + r$, where r is 1 or 5 or 7 or 11. (These are the only possibilities for a prime, except when p is 2 or 3, which is excluded.) Then the inequality is $2k + \frac{1}{6}r < x < 4k + \frac{1}{3}r$. Again we can ignore the even numbers $2k$ and $4k$, and we are left with $\frac{1}{6}r < x < \frac{1}{3}r$. This has no solution if r is 1, one solution if r is 5 or 7, two solutions if r is 11. Hence 3 *is a quadratic residue for primes of the form* $12k \pm 1$, *and a quadratic non-residue for primes of the form* $12k \pm 5$.

5. *The law of reciprocity*

We have just proved that the quadratic character of 2 (mod p) depends only on the remainder r when p is expressed in the form $8k + r$, and that the quadratic character of 3 (mod p) depends only on the remainder r' when p is expressed in the form $12k + r'$. Moreover, in the former case the result is the same for r and for $8 - r$, and in the latter case it is the same for r' and $12 - r'$.

On the basis of extensive numerical evidence, Euler came to the conclusion that a similar state of affairs holds generally, though he was unable to prove it. Let a be any natural number, and express p as $4ak + r$, where $0 < r < 4a$. Then Euler conjectured that *the quadratic character of a (mod p) is the same for all primes p for which r has the same value, and moreover is the same for r and for $4a - r$.* This result is equivalent to the law of quadratic reciprocity, which we shall formulate later in this section. Legendre gave an incomplete proof, and the first complete proof (a very difficult one) was that of Gauss, who discovered the law for himself at the age of nineteen.

It is possible to prove Euler's conjecture by using Gauss's lemma and following the same line of argument as we used before when a was 2 or 3. We have to consider how many of the numbers

$$a, 2a, 3a, \ldots , Pa, \quad \text{where} \quad P = \tfrac{1}{2}(p-1),$$

lie between $\frac{1}{2}p$ and p, or between $\frac{3}{2}p$ and $2p$, and so on. Since Pa is the largest multiple of a that is less than $\frac{1}{2}pa$, the last interval in the series which we have to consider is the interval from $(b-\frac{1}{2})p$ to bp, where b is $\frac{1}{2}a$ or $\frac{1}{2}(a-1)$, whichever is an integer. Thus we have to consider how many multiples of a lie in the intervals

$$(\tfrac{1}{2}p, p), (\tfrac{3}{2}p, 2p), \ldots, ((b-\tfrac{1}{2})p, bp).$$

None of the numbers occurring here is itself a multiple of a, and so no question arises as to whether any of the endpoints of the intervals is to be counted or not.

Dividing throughout by a, we see that the number in question is the total number of integers in all the intervals

$$\left(\frac{p}{2a}, \frac{p}{a}\right), \left(\frac{3p}{2a}, \frac{2p}{a}\right), \ldots, \left(\frac{(2b-1)p}{2a}, \frac{bp}{a}\right).$$

Now write $p=4ak+r$. Since the denominators are all a or $2a$, we can see without any calculation that the effect of replacing p by $4ak+r$ is the same as that of replacing p by r, except that certain even numbers are added to the endpoints of the various intervals. As before, we can ignore these even numbers. It follows that if v is the total number of integers in all the intervals

$$\left(\frac{r}{2a}, \frac{r}{a}\right), \left(\frac{3r}{2a}, \frac{2r}{a}\right), \ldots, \left(\frac{(2b-1)r}{2a}, \frac{br}{a}\right) \tag{10}$$

then a is a quadratic residue or non-residue (mod p) according as v is even or odd. The number v depends only on r, and not on the particular prime p which leaves the remainder r when divided by $4a$.

This proves the main part of Euler's conjecture. Now consider the effect of changing r into $4a-r$. This changes the series of intervals (10) into the series

$$\left(2-\frac{r}{2a}, 4-\frac{r}{a}\right), \left(6-\frac{3r}{2a}, 8-\frac{2r}{a}\right), \ldots \tag{11}$$
$$\left(4b-2-\frac{(2b-1)r}{2a}, 4b-\frac{br}{a}\right).$$

If v' denotes the total number of integers in these intervals, we have to prove that v and v' are of the same parity. In fact, a little consideration shows that the interval $\left(2-\frac{r}{2a}, 4-\frac{r}{a}\right)$ is equivalent to the interval $\left(\dfrac{r}{2a}, \dfrac{r}{a}\right)$, as far as the parity of the number of

integers in it is concerned. For if we subtract both numbers from 4,

the former interval becomes $\left(\dfrac{r}{a}, 2+\dfrac{r}{2a}\right)$. Together with the latter

interval $\left(\dfrac{r}{2a}, \dfrac{r}{a}\right)$, this just makes up an interval of length 2, and

such an interval contains exactly 2 integers. A similar consideration applies to the other intervals in the two series (10) and (11),
and it follows that $v+v'$ is even, which proves the result.

The *law of quadratic reciprocity* was first clearly formulated by
Legendre in 1785. It relates to two different primes p and q, and
gives a rule for the quadratic character of p (mod q) in terms of the
quadratic character of q (mod p). The rule is that *the characters are
the same unless p and q are both of the form* $4k+3$, *in which case
they are opposite*. This can be expressed symbolically by the
formula

$$\left(\frac{p}{q}\right)\left(\frac{q}{p}\right) = (-1)^{\frac{p-1}{2}\cdot\frac{q-1}{2}}. \tag{12}$$

The exponent of -1 on the right is even unless p and q are both of
the form $4k+3$ in which case it is odd. We shall deduce the law of
reciprocity from the results just proved about the quadratic
character of a fixed number a to various prime moduli.

Suppose first that $p\equiv q$ (mod 4). We can suppose without loss of
generality that $p>q$, and we write $p-q=4a$. Then, since $p=4a+q$,
we have

$$\left(\frac{p}{q}\right) = \left(\frac{4a+q}{q}\right) = \left(\frac{4a}{q}\right) = \left(\frac{a}{q}\right).$$

Similarly

$$\left(\frac{q}{p}\right) = \left(\frac{p-4a}{p}\right) = \left(\frac{-4a}{p}\right) = \left(\frac{-1}{p}\right)\left(\frac{a}{p}\right).$$

Now $\left(\dfrac{a}{p}\right)$ and $\left(\dfrac{a}{q}\right)$ are the same, because p and q leave the same

remainder on division by $4a$. Hence

$$\left(\frac{p}{q}\right)\left(\frac{q}{p}\right) = \left(\frac{-1}{p}\right),$$

and this is 1 if p and q are both of the form $4k+1$, and -1 if they are both of the form $4k+3$.

Suppose next that $p \not\equiv q$ (mod 4); in this case $p \equiv -q$ (mod 4). Put $p+q=4a$. Then, in the same way as before, we obtain

$$\left(\frac{p}{q}\right) = \left(\frac{4a-q}{q}\right) = \left(\frac{4a}{q}\right) = \left(\frac{a}{q}\right),$$

and similarly $\left(\frac{q}{p}\right) = \left(\frac{a}{p}\right)$. Again $\left(\frac{a}{p}\right)$ and $\left(\frac{a}{q}\right)$ are the same, since

p and q leave opposite remainders on division by $4a$. This completes the proof of the law of reciprocity.

The law of quadratic reciprocity is one of the most famous theorems in the whole of the theory of numbers. It reveals a simple and striking relationship between the solubility of the congruences $x^2 \equiv q$ (mod p) and $x^2 \equiv p$ (mod q), a relationship which is by no means obvious. The desire to find what lies behind the law has been an important factor in the work of many mathematicians, and has led to far-reaching discoveries. The first rigorous proof, given by Gauss in his *Disquisitiones*, was by induction on the two primes p and q, and such a proof is necessarily both difficult and unsatisfying. Gauss himself gave altogether seven proofs, based on widely different methods and exhibiting the connection between the law of reciprocity and various other arithmetical theories.

The law of reciprocity enables one to calculate the value of $(a \mid p)$, in any numerical case, without referring to the solubility of congruences. As an example, we calculate $(34 \mid 97)$. The first step is to factorize 34 as 2×17. Since 97 is a prime of the form $8k+1$, we have $(2 \mid 97)=1$, and so $(34 \mid 97)=(17 \mid 97)$. Since 17 and 97 are primes, not both of the form $4k+3$, the law of reciprocity tells us that $(17 \mid 97)=(97 \mid 17)$, or $(12 \mid 17)$ since $97 \equiv 12$ (mod 17). Now $(12 \mid 17)=(3 \mid 17)=(17 \mid 3)$, on applying the law of quadratic reciprocity again. Since $17 \equiv -1$ (mod 3), the value of the symbol is $(-1 \mid 3)$, or -1.

There is no such simple law as that of quadratic reciprocity for cubic or higher power residues. But we may mention briefly one result of Gauss concerning fourth power residues. First we must recall that, by the results of §1, the theory of fourth power residues is significant only for primes of the form $4n+1$. For if p is of the form $4n+3$, the highest common factor of 4 and $p-1$ is 2, that is $K=2$ in the notation of §1, and therefore in this case the fourth power residues are just the same as the quadratic residues. But if p is of the form $4n+1$, half the quadratic residues are fourth power residues (namely those whose indices are divisible by 4), and the other half together with all the quadratic non-residues are fourth power non-residues. The result of Gauss is that the number 2 is a fourth power residue (mod p) if and only if the prime p is representable as x^2+64y^2. It may be remarked that the prime p, being of the form $4n+1$, is necessarily representable as a^2+b^2 (as we shall prove in Chapter V), and obviously one of a and b must be odd and the other even. So Gauss's condition is that the even one of a and b must be divisible by 8. For example, 2 is a fourth power residue (mod 73), since $73=3^2+64$.

6. The distribution of the quadratic residues

We now return to questions connected with the quadratic residues and non-residues to a single prime modulus p. We know that half of the numbers

$$1, 2, \ldots, p-1$$

are quadratic residues, and the other half non-residues. A few trials will soon suggest that if p is a large prime, the residues and non-residues have a distribution which is fairly random. It is, of course, subject to the laws we know; for example the multiplicative law and the fact that any perfect square is always a quadratic residue.

There are various questions which may be proposed to test the random character of the distribution. We may ask, for example, how the residues and non-residues are distributed in a sub-interval of the interval from 0 to p. Suppose that α and β are two fixed proper fractions; is it true when p is large that about half the numbers between αp and βp are quadratic residues? If so, we may express this by saying that the quadratic residues are *equally* distributed. This proposition is in fact true, but there does not seem to be any very elementary proof of it.

An easier question, which was answered by Gauss, concerns the characters of consecutive numbers. If n and $n+1$ are two consecutive numbers in the series $1, 2, \ldots, p-1$, how often does it happen that they have prescribed characters? The possible characters for a pair of numbers are RR, RN, NR, NN. If we think that the quadratic residues and non-residues are distributed randomly, we may expect that each of the four types will occur about equally often.

This is in fact the case, as is not difficult to prove. Let us denote by (RR), and so on, the number of pairs, n, $n+1$ with prescribed characters. Plainly $(RR)+(RN)$ is the number of pairs for which n is a quadratic residue. Here n takes the values $1, 2, \ldots, p-2$. The total number of quadratic residues among $1, 2, \ldots, p-1$ is $\frac{1}{2}(p-1)$, and the character of the number $p-1$, or -1, is $(-1)^{\frac{1}{2}(p-1)}$. Hence

$$(RR)+(RN)=\tfrac{1}{2}(p-2-\varepsilon), \tag{13}$$

where $\varepsilon=(-1)^{\frac{1}{2}(p-1)}$. Similarly we find that

$$(NR)+(NN)=\tfrac{1}{2}(p-2+\varepsilon), \tag{14}$$

$$(RR)+(NR)=\tfrac{1}{2}(p-1)-1, \tag{15}$$

$$(RN)+(NN)=\tfrac{1}{2}(p-1). \tag{16}$$

These are four relations for the four unknowns, but they are not independent, because on adding the first two we get the same result as on adding the last two. So we need another relation in order to determine the four unknowns.

Consider the product of the Legendre symbols $(n \mid p)$ and $(n+1 \mid p)$. This is $+1$ in the cases RR and NN, and -1 in the cases RN and NR. Hence

$$(RR)+(NN)-(RN)-(NR)$$

is equal to the sum of all the Legendre symbols

$$\left(\frac{n(n+1)}{p}\right),$$

where n takes the values $1, 2, \ldots, p-2$. Any integer n in this set has a reciprocal (mod p), which we shall denote by m. Now $n(n+1)\equiv n^2(1+m)$ (mod p), hence

$$\left(\frac{n(n+1)}{p}\right) = \left(\frac{1+m}{p}\right).$$

As n takes the values $1, 2, \ldots, p-2$, i.e. all the values from 1 to $p-1$ except $p-1$, its reciprocal m also takes all values from 1 to $p-1$ except $p-1$. Hence $1+m$ takes all values from 2 to $p-1$. The sum of the Legendre symbols of these numbers is

$$\left(\frac{2}{p}\right) + \left(\frac{3}{p}\right) + \ldots + \left(\frac{p-1}{p}\right).$$

Now $\left(\dfrac{1}{p}\right) + \left(\dfrac{2}{p}\right) + \left(\dfrac{3}{p}\right) + \ldots + \left(\dfrac{p-1}{p}\right) = 0,$

since there are as many residues as non-residues. Hence the sum we are interested in has the value $-(1 \mid p)$, or -1. Thus

$$(RR)+(NN)-(RN)-(NR)=-1. \tag{17}$$

This relation, combined by addition and subtraction with the earlier relations, gives us the values of (RR), etc. If we add (17) to (13) and (14), we obtain

$$(RR)+(NN)=\tfrac{1}{2}(p-3).$$

On the other hand, subtracting (14) from (15) gives

$$(RR)-(NN) = -\tfrac{1}{2}(1+\varepsilon).$$

Hence $(RR)=\tfrac{1}{4}(p-4-\varepsilon),$

and similarly we get the other three numbers. From the results we find that the value of each of the four numbers (RR), etc., is between $\tfrac{1}{4}(p-5)$ and $\tfrac{1}{4}(p+1)$. So the assertion that they are all about $\tfrac{1}{4}p$ for large p is amply justified.

The important step in the proof was the evaluation of the sum of the Legendre symbols $\left(\dfrac{n(n+1)}{p}\right)$. If we make the convention that $(0 \mid p)=0$, we can allow n to take a complete set of values $0, 1, \ldots ,$ $p-1$ instead of only the values $1, 2, \ldots , p-2$, without altering the sum. Hence the result can be expressed in the form

$$\Sigma \left(\dfrac{n(n+1)}{p}\right) = -1, \tag{18}$$

where the symbol Σ denotes summation for n over a complete set of residues (mod p). This result can be shown to hold more generally for any sum

$$\Sigma \left(\dfrac{n^2+bn+c}{p}\right),$$

formed with a quadratic polynomial with highest coefficient 1; though not by the method used above. There is an obvious exception, of course, if the polynomial is a perfect square. Similar questions for polynomials of higher degree have been deeply investigated during the last fifty years or so. Hasse showed in 1934, by very difficult and advanced methods, that any cubic sum

$$\dot{\Sigma} \left(\frac{an^3 + bn^2 + cn + d}{p} \right)$$

has a value between $-2\sqrt{p}$ and $2\sqrt{p}$. This result was later generalized, with far-reaching consequences, by A. Weil.

NOTES

§1. There is another proof of the existence of a primitive root, due to Gauss. But I have preferred Legendre's proof as being of a more constructive nature.

In accordance with the theorem of Fermat and Euler (II. 3), a number is considered to be a primitive root to a general modulus m if its order is exactly $\phi(m)$. It was proved by Gauss that primitive roots exist for the moduli 2, 4, p^n, $2p^n$, where p is any prime greater than 2 and n is any natural number, but for no other moduli.

§2. There is a table of indices for primes up to 97 in Uspensky and Heaslet.

§3. One can prove the multiplicative property and Euler's criterion directly from the definition of a quadratic residue, without using indices, but the proofs are less illuminating.

§4. In §3 we gave Lagrange's explicit construction for the solution of $x^2 \equiv -1 \pmod{p}$ when p is a prime of the form $4k + 1$. There is the similar problem of giving an explicit construction for the solution of $x^2 \equiv 2 \pmod{p}$ when p is a prime of the form $8k + 1$ or $8k - 1$. In the second of these two cases there is a simple answer, namely $x = 2^{2k}$, since $2^{4k-1} = 2^{\frac{1}{2}(p-1)} \equiv 1 \pmod{p}$ by Euler's criterion. No simple answer has been given in the case $p = 8k + 1$.

§5. In adopting this approach to the law of reciprocity, I am following Scholz in his *Einführung in die Zahlentheorie*.

§6. The fact that the quadratic residues and non-residues are equally distributed follows from an important inequality discovered by Pólya in 1917 and independently by Vinogradov in 1918. It is that the sum of the Legendre symbols $(n|p)$ over *any* range of consecutive integers n is in absolute value less than $Cp^{\frac{1}{2}} \log p$, where C is a certain constant. Since $p^{\frac{1}{2}} \log p$ is small compared with p when p is large, it follows that there are almost as many residues as non-residues in an interval from αp to βp, where α and β are fixed and p is large. For further and deeper results on the distribution of quadratic residues and non-residues, see D. A. Burgess, *Mathematika*, 4 (1957), 106–112, or Gelfond and Linnik, ch. 9. For an elementary exposition of Hasse's proof, due to Manin, see Gelfond and Linnik, ch. 10.

IV

CONTINUED FRACTIONS

1. Introduction

In I.6 we discussed Euclid's algorithm for finding the highest common factor of two given natural numbers. There is another way of expressing the algorithm, the effect of which is to represent the quotient of the two numbers as a continued fraction. The method will become clear from a numerical example.

Let us apply Euclid's algorithm to the numbers 67 and 24. The successive steps are:

$$67 = 2 \times 24 + 19,$$
$$24 = 1 \times 19 + 5,$$
$$19 = 3 \times 5 + 4,$$
$$5 = 1 \times 4 + 1.$$

The last remainder is 1, as we know must be the case because the numbers 67 and 24 are relatively prime. We now write each of the equations in fractional form:

$$\tfrac{67}{24} = 2 + \tfrac{19}{24},$$
$$\tfrac{24}{19} = 1 + \tfrac{5}{19},$$
$$\tfrac{19}{5} = 3 + \tfrac{4}{5},$$
$$\tfrac{5}{4} = 1 + \tfrac{1}{4}.$$

The last fraction in each of these equations is the reciprocal of the first fraction in the following equation. We can therefore eliminate all the intermediate fractions, and express the original fraction $\tfrac{67}{24}$ in the form

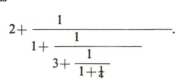

Such an expression is called a *continued fraction*. For convenience of writing and printing, one adopts the form

$$2 + \cfrac{1}{1+} \cfrac{1}{3+} \cfrac{1}{1+} \cfrac{1}{4}.$$

The numbers 2, 1, 3, 1, 4 here are called the *terms* of the continued fraction, or the *partial quotients*, since they are the partial quotients in the successive steps of Euclid's algorithm applied to the numerator and denominator of the original fraction. The *complete quotients* are the numbers $\frac{67}{24}$, $\frac{24}{19}$, $\frac{19}{5}$, $\frac{5}{4}$ themselves. Each of these has a continued fraction which is derived from that above by starting at a later term, e.g.

$$\frac{24}{19} = 1 + \cfrac{1}{3+} \cfrac{1}{1+} \cfrac{1}{4}, \qquad \frac{19}{5} = 3 + \cfrac{1}{1+} \cfrac{1}{4}.$$

It is plain from the above example, and from what we know about Euclid's algorithm, that each rational number $\dfrac{a}{b}$ greater than 1 can be represented by a continued fraction:

$$\frac{a}{b} = q + \cfrac{1}{r+} \cfrac{1}{s+} \dots \cfrac{1}{w},$$

whose terms $q, r, s, \dots, w$ are natural numbers. The last term, w above, must be greater than 1, because it is the last quotient in Euclid's algorithm.

It is very easy to prove that there is only one representation of a given rational number as a continued fraction. For suppose that

$$\frac{a}{b} = q + \cfrac{1}{r+} \cfrac{1}{s+} \dots = q' + \cfrac{1}{r'+} \cfrac{1}{s'+} \dots,$$

where $q', r', s', \dots$ are also natural numbers, the last of which is greater than 1. The amount added to q on the left is less than 1, and so is the amount added to q' on the right. So q and q' are both equal to the integral part of the rational number $\dfrac{a}{b}$, and are the same. Cancelling q against q' and inverting, we get

$$r + \cfrac{1}{s+} \cdots = r' + \cfrac{1}{s'+} \cdots .$$

The same argument proves that $r=r'$, and so on generally.

Before going further, the reader who is unacquainted with continued fractions should practise developing a few simple rational numbers. Examples are:

$$\frac{17}{11} = 1 + \cfrac{1}{1+} \cfrac{1}{1+} \frac{1}{5}, \qquad \frac{11}{31} = \cfrac{1}{2+} \cfrac{1}{1+} \cfrac{1}{4+} \frac{1}{2}.$$

Where the rational number is less than 1, as in the second example, the first partial quotient is 0 and is omitted.

2. *The general continued fraction*

Continued fractions are of great service in the theory of numbers; by using them one can often give an explicit construction for the solution of a problem, where other methods would prove only that a solution exists.

We write the general continued fraction in the form

$$q_0 + \cfrac{1}{q_1+} \cfrac{1}{q_2+} \cdots \cfrac{1}{q_n}. \qquad (1)$$

Before we can usefully investigate the arithmetical properties of continued fractions we need some purely algebraic relations. These relations are identities, whose validity does not depend on the nature of the terms $q_0, q_1, \dots, q_n$. For the time being, therefore, we treat the terms as variables, not necessarily natural numbers.

If we work out the continued fraction (1) in stages, we shall obviously end with an expression for it as the quotient of two sums, each sum comprising various products formed with $q_0, q_1, \dots, q_n$. If n is 1, we have

$$q_0 + \frac{1}{q_1} = \frac{q_0 q_1 + 1}{q_1}.$$

If n is 2, we have

$$q_0 + \cfrac{1}{q_1+} \cfrac{1}{q_2} = q_0 + \frac{q_2}{q_1 q_2 + 1} = \frac{q_0 q_1 q_2 + q_0 + q_2}{q_1 q_2 + 1},$$

where in the intermediate step we have quoted the value of $q_1 + \cfrac{1}{q_2}$

from the previous calculation, putting q_1 and q_2 in place of q_0 and q_1. Similarly, when n is 3, we have

$$q_0 + \frac{1}{q_1+} \frac{1}{q_2+} \frac{1}{q_3} = q_0 + \frac{q_2q_3+1}{q_1q_2q_3+q_1+q_3}$$

$$= \frac{q_0q_1q_2q_3+q_0q_1+q_0q_3+q_2q_3+1}{q_1q_2q_3+q_1+q_3}. \tag{2}$$

Here again we have used the result of the previous step.

It is plain that we can build up the general continued fraction by going on in this way. We shall denote the numerator of the continued fraction (1), when evaluated in this way, by

$$[q_0, q_1, ..., q_n].$$

Thus $$[q_0]=q_0, \quad [q_0, q_1]=q_0q_1+1,$$

$$[q_0, q_1, q_2]=q_0q_1q_2+q_0+q_2,$$

$$[q_0, q_1, q_2, q_3]=q_0q_1q_2q_3+q_0q_1+q_0q_3+q_2q_3+1,$$

and so on. It will be seen that in the cases worked out above, the denominator of the expression obtained for the continued fraction is

$$[q_1, q_2, ... , q_n].$$

This is true generally. For if we look at the third stage (which is quite typical) in (2) above, the denominator of the answer comes from the numerator of $q_1 + \dfrac{1}{q_2+} \dfrac{1}{q_3}$, and so has the value

$$[q_1, q_2, q_3].$$

The general continued fraction therefore has the value

$$q_0 + \frac{1}{q_1+} \cdots \frac{1}{q_n} = \frac{[q_0, q_1, ... , q_n]}{[q_1, q_2, ... , q_n]}. \tag{3}$$

It is plain from the calculation in (2) how the function $[q_0, q_1, q_2, q_3]$ is built up out of $[q_1, q_2, q_3]$ and $[q_2, q_3]$. That calculation shows, namely, that

$$[q_0, q_1, q_2, q_3]=q_0[q_1, q_2, q_3]+[q_2, q_3].$$

This is obviously typical of the general case, and we have the rule

$$[q_0, q_1, ... , q_n]=q_0[q_1, q_2, ... , q_n]+[q_2, q_3 ... , q_n]. \tag{4}$$

This is a *recurrence relation,* which defines the square-bracket function step by step. As it stands, the formula applies from $n=2$ onwards. It still applies when n is 1, if we give the interpretation 1 to the second square bracket on the right, which in itself is meaningless in this case. With this interpretation, the formula becomes

$$[q_0, q_1]=q_0[q_1]+1=q_0q_1+1,$$

which is correct.

As an illustration, we can apply the rule to the last example mentioned at the end of §1. We have

$$[4, 2]=4\times2+1=9,$$
$$[1, 4, 2]=1\times[4, 2]+[2]=9+2=11,$$
$$[2, 1, 4, 2]=2[1, 4, 2]+[4, 2]=2\times11+9=31.$$

Thus

$$2 + \frac{1}{1+} \frac{1}{4+} \frac{1}{2} = \frac{[2, 1, 4, 2]}{[1, 4, 2]} = \frac{31}{11}.$$

One word of caution is necessary. We have seen that we can express the general continued fraction in the form (3), where the two square brackets are certain sums of products of the variables $q_0, q_1, \dots, q_n$. We have *not* proved that nothing can be cancelled from the numerator and denominator in this representation. This is actually true, and it is true in two senses, one algebraical and one arithmetical. In the former sense, the numerator and denominator are polynomials in the variables $q_0, q_1, \dots, q_n$, and it can be proved that these polynomials are irreducible, that is they cannot be factorized into other polynomials. In the latter sense, if $q_0, q_1, \dots,$ q_n are integers, the numerator and denominator are integers and are always relatively prime. This second fact will be proved in §4. The first fact is even more easily proved, but is of no interest from the point of view of the theory of numbers.

3. *Euler's rule*

We have seen that $[q_0, \dots, q_n]$ is the sum of certain products formed out of the terms $q_0, q_1, \dots, q_n$. Which products are these? The answer was given by Euler, who was the first to give a general account of continued fractions. *First take the product of all the terms. Then take every product that can be obtained by omitting any pair of consecutive terms. Then take every product that can be obtained by omitting any two separate pairs of consecutive terms, and so on.* The sum of all such products gives the value of

$$[q_0, q_1, \cdots , q_n].$$

It is to be understood that if $n+1$ is even, we end by including the empty product which is got by omitting all the terms, giving this the conventional value 1. An example of Euler's rule is:

$$[q_0, q_1, q_2, q_3] = q_0 q_1 q_2 q_3 + q_2 q_3 + q_0 q_3 + q_0 q_1 + 1.$$

Here we have taken first the product of all the terms, then the product with the pair q_0, q_1 omitted, then with the pair q_1, q_2 omitted, then with the pair q_2, q_3 omitted, and finally the empty product with both the pairs q_0, q_1 and q_2, q_3 omitted. Another example, with one more term, is:

$$\begin{aligned} [q_0, q_1, q_2, q_3, q_4] = {} & q_0 q_1 q_2 q_3 q_4 \\ & + q_2 q_3 q_4 + q_0 q_3 q_4 + q_0 q_1 q_4 + q_0 q_1 q_2 \\ & + q_4 + q_2 + q_0. \end{aligned}$$

In the second line we have written all the products with one pair of consecutive terms omitted, and on the last line the results of omitting two separate pairs, e.g. omitting q_0, q_1 and q_2, q_3 gives q_4.

Having verified that the rule is correct for the first few of the square-bracket functions, we can prove it generally by induction, using the recurrence relation (4). Assuming the rule holds for the two square-bracket functions on the right of (4), we have to prove that it holds for the one on the left. The expression $[q_2, \ldots , q_n]$ represents the sum of all those products formed from $q_0, q_1, \ldots , q_n$ in which the pair q_0, q_1 is omitted. Now $q_0[q_1, \ldots , q_n]$ represents precisely the sum of all those products formed from $q_0, q_1, \ldots , q_n$ in which the pair q_0, q_1 is *not* one of those omitted; for all such products must contain q_0, and when this factor is removed we are left with the sum of all products of $q_1, \ldots , q_n$ from which any separate pairs of consecutive terms are omitted. Together, we get the appropriate sum of products of $q_0, q_1, \ldots , q_n$, and so the rule holds for the function $[q_0, q_1, \ldots , q_n]$. This proves the rule generally, by induction on the number of variables.

One immediate deduction from Euler's rule is that *the value of* $[q_0, q_1, \ldots , q_n]$ *is unchanged if the terms are written in the opposite order*:

$$[q_0, q_1, \ldots , q_n] = [q_n, q_{n-1}, \ldots , q_0].$$

For example,

$$[2, 4, 1, 2] = [2, 1, 4, 2] = 31.$$

It follows from this fact that besides the recurrence relation (4) there is a similar relation which expresses $[q_0, q_1, \ldots, q_n]$ in terms of the similar functions with the last term or last two terms omitted. This relation is

$$[q_0, q_1, \ldots, q_n] = q_n[q_0, q_1, \ldots, q_{n-1}] + [q_0, q_1, \ldots, q_{n-2}]. \quad (5)$$

This is equivalent to (4), because if we write the terms in the opposite order it becomes

$$[q_n, q_{n-1}, \ldots, q_0] = q_n[q_{n-1}, \ldots, q_0] + [q_{n-2}, \ldots, q_0],$$

and this is merely a restatement of (4) with different symbols.

The recurrence relation (5) is more convenient than (4) for most purposes. We are more commonly concerned with adding terms at the end of a continued fraction than with adding terms at the beginning, and (5) enables us to investigate what happens when this is done.

4. *The convergents to a continued fraction*

Let

$$q_0 + \frac{1}{q_1+} \cdots \frac{1}{q_n} \quad (6)$$

be any continued fraction. We shall suppose throughout this section that the terms $q_0, q_1, \ldots, q_n$ are natural numbers. The various continued fractions

$$q_0, \quad q_0 + \frac{1}{q_1}, \quad q_0 + \frac{1}{q_1+} \frac{1}{q_2}, \quad \ldots,$$

obtained by stopping at an earlier term than q_n, are called the *convergents* to the continued fraction. The reason why this name is appropriate will become clear later.

The value of the general convergent, obtained by stopping at q_m, say, is

$$q_0 + \frac{1}{q_1+} \cdots \frac{1}{q_m} = \frac{[q_0, \ldots, q_m]}{[q_1, \ldots, q_m]}.$$

In order to have a simpler notation, we put

$$A_m = [q_0, \ldots, q_m], \quad B_m = [q_1, \ldots, q_m], \quad (7)$$

so that the above convergent is $\dfrac{A_m}{B_m}$. The first convergent is

$\frac{A_0}{B_0} = \frac{q_0}{1}$. The last is $\frac{A_n}{B_n}$, which is the value of the continued fraction itself. The numbers A_0, B_0, A_1, B_1, ... are all natural numbers, being sums of products formed out of the q's in accordance with Euler's rule.

The recurrence relation (5) now takes the simple form

$$A_m = q_m A_{m-1} + A_{m-2}. \qquad (8)$$

The same recurrence relation, with q_0 omitted, tells us that

$$B_m = q_m B_{m-1} + B_{m-2}. \qquad (9)$$

Thus the numerators and denominators of the convergents are formed by the same general rules. These rules are very convenient for purposes of numerical calculation; we can write down the first two convergents by inspection, and the subsequent ones by applying the rule. For example, the continued fraction for $\frac{42}{31}$ is

$$1 + \frac{1}{2+} \frac{1}{1+} \frac{1}{4+} \frac{1}{2}.$$

The first two convergents are obviously $\frac{1}{1}$ and $\frac{3}{2}$. Since the next partial quotient is 1, the next convergent is $\frac{3+1}{2+1} = \frac{4}{3}$. The next partial quotient is 4, so the next convergent is

$$\frac{4 \times 4 + 3}{4 \times 3 + 2} = \frac{19}{14}.$$

The final partial quotient is 2, and the final convergent is

$$\frac{2 \times 19 + 4}{2 \times 14 + 3} = \frac{42}{31},$$

which is, of course, the original number.

There is a simple relation satisfied by any two consecutive convergents, which is of the greatest importance. It is that

$$A_m B_{m-1} - B_m A_{m-1} = (-1)^{m-1}. \qquad (10)$$

For example, if m is 1, we have

$$A_0 = q_0, \; B_0 = 1, \; A_1 = q_0 q_1 + 1, \; B_1 = q_1,$$

and so

$$A_1 B_0 - B_1 A_0 = (q_0 q_1 + 1) - q_0 q_1 = 1. \qquad (11)$$

To prove (10) generally, we substitute for A_m and B_m from the recurrence relations (8) and (9). This gives

$$A_m B_{m-1} - B_m A_{m-1}$$
$$= (q_m A_{m-1} + A_{m-2}) B_{m-1} - (q_m B_{m-1} + B_{m-2}) A_{m-1}$$
$$= - (A_{m-1} B_{m-2} - B_{m-1} A_{m-2}).$$

Consequently the expression on the left of (10), say Δ_m, has the property that $\Delta_m = -\Delta_{m-1}$. Hence

$$\Delta_m = -\Delta_{m-1} = +\Delta_{m-2} = \ldots = \pm\Delta_1,$$

and the sign at the end is $+1$ if m is odd and -1 if m is even, so that it can be represented by $(-1)^{m-1}$. Since $\Delta_1 = 1$ by (11), the general result (10) follows.

One immediate consequence of (10) is that A_m *and* B_m *are always relatively prime*, for any common factor would have to be a factor of 1. Thus the fraction $\dfrac{A_m}{B_m}$, representing the general convergent, is in its lowest terms. In particular, taking m to be n, this is true of the earlier formula (3) for the value of a general continued fraction. Thus we have now proved the statement made at the end of §2.

If we develop a rational number $\dfrac{a}{b}$ into a continued fraction, the convergents to that continued fraction constitute a sequence of rational numbers, the last of which is $\dfrac{a}{b}$ itself. What relations of magnitude are there between these numbers and $\dfrac{a}{b}$ itself? It is quite easy to prove that *the convergents are alternately less than, and greater than, the final value* $\dfrac{a}{b}$. To see this, write the relation (10) in the form

$$\frac{A_m}{B_m} - \frac{A_{m-1}}{B_{m-1}} = \frac{(-1)^{m-1}}{B_{m-1} B_m}. \tag{12}$$

This shows that the difference on the left is positive if m is odd and negative if m is even. Also, since the numbers $B_0, B_1, B_2, \ldots$ increase steadily, the difference in (12) decreases steadily as m increases. Thus $\dfrac{A_1}{B_1}$ is greater than $\dfrac{A_0}{B_0}$, and $\dfrac{A_2}{B_2}$ is less than $\dfrac{A_1}{B_1}$ but greater than $\dfrac{A_0}{B_0}$, and $\dfrac{A_3}{B_3}$ is greater than $\dfrac{A_2}{B_2}$ but less than $\dfrac{A_1}{B_1}$, and so

on. Since we end with $\dfrac{A_n}{B_n}=\dfrac{a}{b}$, it follows that all the even conver-

gents $\dfrac{A_0}{B_0}$, $\dfrac{A_2}{B_2}$, ... are less than $\dfrac{a}{b}$, and all the odd convergents are

greater than $\dfrac{a}{b}$.

It can be proved that *each convergent is nearer to the final value*
$\dfrac{a}{b}$ *than the preceding convergent.* The proof is not difficult, but we
omit it here. Another interesting fact is that the convergents are
the 'best possible' approximations to $\dfrac{a}{b}$ by fractions with specified
complexity. We measure the complexity of a fraction by the size of
its denominator. Thus any fraction which is nearer to $\dfrac{a}{b}$ than a
particular convergent $\dfrac{A_m}{B_m}$ must have a denominator which is
greater than B_m.

To illustrate these properties of the convergents, take the continued fraction for $\frac{42}{31}$, mentioned earlier in this section. The
successive convergents are $\frac{1}{1}$, $\frac{3}{2}$, $\frac{4}{3}$, $\frac{19}{14}$, $\frac{42}{31}$. When expressed as
decimals, these numbers are

$$1,\ 1.5,\ 1.333...\ ,\ 1.3571...\ ,\ 1.3548...,$$

and we see that they are alternately less than and greater than the
final number, and are successively nearer to it.

5. The equation $ax-by=1$

It was proved in I.8 that if a and b are any two relatively prime
natural numbers, then it is possible to find natural numbers x and
y to satisfy the equation $ax-by=1$. The process for converting
$\dfrac{a}{b}$ into a continued fraction provides an explicit construction for
two such numbers. Suppose the continued fraction is

$$\frac{a}{b} = q_0 + \frac{1}{q_1+} \cdots \frac{1}{q_n}.$$

The last convergent $\dfrac{A_n}{B_n}$ is $\dfrac{a}{b}$ itself. The preceding convergent

$\dfrac{A_{n-1}}{B_{n-1}}$ satisfies

$$A_n B_{n-1} - B_n A_{n-1} = (-1)^{n-1}, \quad \text{or} \quad aB_{n-1} - bA_{n-1} = (-1)^{n-1},$$

by (10) of the preceding section. Hence, if we take $x = B_{n-1}$ and $y = A_{n-1}$, we have a solution in natural numbers of the equation $ax - by = (-1)^{n-1}$. If n is odd, this is the equation proposed. If n is even, so that $(-1)^{n-1} = -1$, we can still solve the equation with $+1$, by either of two methods (which are in fact the same). One method is to take $x = b - B_{n-1}$ and $y = a - A_{n-1}$; then

$$ax - by = a(b - B_{n-1}) - b(a - A_{n-1}) = -aB_{n-1} + bA_{n-1} = 1.$$

The other method is to modify the continued fraction by replacing the last term q_n by $(q_n - 1) + \dfrac{1}{1}$. The new continued fraction has one more term than the old, and so its penultimate convergent provides a solution of the equation with $+1$ on the right. In fact, this will give the same solution as the other method.

To take a simple numerical example, suppose we wish to find natural numbers x and y which satisfy

$$61x - 48y = 1.$$

The continued fraction for $\frac{61}{48}$ is

$$\frac{61}{48} = 1 + \frac{1}{3+} \frac{1}{1+} \frac{1}{2+} \frac{1}{4}.$$

The convergents to it are

$$\tfrac{1}{1}, \tfrac{4}{3}, \tfrac{5}{4}, \tfrac{14}{11}, \tfrac{61}{48}.$$

Since n is 4 in this case, the numbers $x = 11$ and $y = 14$ satisfy the equation $61x - 48y = -1$. To solve the equation proposed, we take $x = 48 - 11 = 37$, $y = 61 - 14 = 47$. Or, alternatively, we modify the continued fraction to

$$1 + \frac{1}{3+} \frac{1}{1+} \frac{1}{2+} \frac{1}{3+} \frac{1}{1}.$$

The convergents are now

$$\tfrac{1}{1}, \tfrac{4}{3}, \tfrac{5}{4}, \tfrac{14}{11}, \tfrac{47}{37}, \tfrac{61}{48},$$

and the penultimate convergent, $\frac{47}{37}$, provides the solution.

It may be noted that this construction provides the least solution of the equation, namely that for which x is less than b and y is less than a. If this solution is denoted by x_0, y_0 then the general solution is given by

$$x = x_0 + bt, \quad y = y_0 + at$$

where t is any integer, positive or zero. Unless t is zero, x is greater than b and y is greater than a.

6. *Infinite continued fractions*

So far, we have been considering the expression of a *rational* number as a continued fraction. It is also possible to represent an *irrational* number by a continued fraction, but in this case the expansion goes on for ever instead of coming to an end.

Let α be any irrational number. Let q_0 be the integral part of α, that is, the greatest integer which is less than α. Then $\alpha = q_0 + \alpha'$, where α' is the fractional part of α, and satisfies $0 < \alpha' < 1$. Put $\alpha' = \dfrac{1}{\alpha_1}$; then

$$\alpha = q_0 + \frac{1}{\alpha_1}, \quad \text{where} \quad \alpha_1 > 1.$$

Plainly α_1 is again irrational, for if it were rational then α would itself be rational. Now repeat the operation on α_1, expressing it as

$$\alpha_1 = q_1 + \frac{1}{\alpha_2}, \quad \text{where} \quad \alpha_2 > 1.$$

We can continue this process indefinitely. Having reached α_n, itself an irrational number greater than 1, we can express it as

$$\alpha_n = q_n + \frac{1}{\alpha_{n+1}}, \quad \text{where} \quad \alpha_{n+1} > 1,$$

and q_n is a natural number. If we combine all the equations up to this one, we obtain for α the expression

$$\alpha = q_0 + \frac{1}{q_1 +} \cdots \frac{1}{q_n +} \frac{1}{\alpha_{n+1}}. \tag{13}$$

All the numbers $q_1, \ldots, q_n$ are natural numbers, and q_0 is an integer which may be positive, negative, or zero. If $\alpha > 1$, then q_0 is positive, and all the terms are natural numbers. The numbers $q_0, q_1, \ldots$ are called, as before, the *terms*, or *partial quotients*, of the continued fraction, and the *complete* quotient corresponding to q_n is α_n, or, what is the same thing, $q_n + \dfrac{1}{\alpha_{n+1}}$. The process can never come to an end, because each complete quotient $\alpha_1, \alpha_2, \ldots$ is an irrational number.

The convergents to the continued fraction are

$$\frac{A_0}{B_0} = q_0, \quad \frac{A_1}{B_1} = q_0 + \frac{1}{q_1}, \quad \frac{A_2}{B_2} = q_0 + \frac{1}{q_1+} \frac{1}{q_2}, \ldots,$$

and they constitute now an infinite sequence of rational numbers. Again they satisfy the recurrence relations (8) and (9), for they are also convergents to the finite continued fraction (13) and all the results proved earlier are applicable. Incidentally, we see now the advantage of not having restricted ourselves, in the initial stages, to continued fractions whose terms are all natural numbers. Had we done so, we should have been precluded from applying our results to the continued fraction (13), as this contains the irrational number α_{n+1}.

The equation (13) allows us to express α in terms of the complete quotient α_{n+1} and the two convergents $\dfrac{A_n}{B_n}$ and $\dfrac{A_{n-1}}{B_{n-1}}$. In fact, using our original notation, (13) means that

$$\alpha = \frac{[q_0, q_1, \ldots, q_n, \alpha_{n+1}]}{[q_1, q_2, \ldots, q_n, \alpha_{n+1}]}.$$

Now, by (5),

$$[q_0, q_1, \ldots, q_n, \alpha_{n+1}] = \alpha_{n+1} [q_0, q_1, \ldots, q_n] + [q_0, q_1, \ldots, q_{n-1}]$$
$$= \alpha_{n+1} A_n + A_{n-1}.$$

Similarly the denominator is $\alpha_{n+1} B_n + B_{n-1}$. Hence

$$\alpha = \frac{\alpha_{n+1} A_n + A_{n-1}}{\alpha_{n+1} B_n + B_{n-1}}. \tag{14}$$

This will be a most serviceable formula throughout the remainder of the chapter.

After realizing that (13) is valid for every n, however large, one is tempted to write simply

$$\alpha = q_0 + \frac{1}{q_1+} \frac{1}{q_2+} \, \dots \tag{15}$$

But before yielding to this very natural temptation, it is advisable to reflect for a moment on the meaning of such a statement. On the face of it, the implication is that we can somehow carry out the infinite number of operations of addition and division which are indicated on the right-hand side, and thereby arrive at a certain number, which is asserted to be α. Now the only way in which one can attach a meaning to the result of carrying out an infinite number of operations is by using the notion of a limit. If we can prove that the sequence of convergents

$$\frac{A_0}{B_0}, \frac{A_1}{B_1}, \frac{A_2}{B_2}, \dots,$$

where

$$\frac{A_n}{B_n} = q_0 + \frac{1}{q_1+} \dots \frac{1}{q_n},$$

has a certain limit as n increases indefinitely, then we can interpret the right-hand side of (15) as meaning the value of this limit. If the limit is in fact α, then (15) will be justified.

It is not difficult to prove that $\dfrac{A_n}{B_n}$ tends to the limit α as n increases indefinitely. The equation (14) gives

$$\alpha - \frac{A_n}{B_n} = \frac{\alpha_{n+1}A_n + A_{n-1}}{\alpha_{n+1}B_n + B_{n-1}} - \frac{A_n}{B_n} = \frac{A_{n-1}B_n - B_{n-1}A_n}{B_n(\alpha_{n+1}B_n + B_{n-1})}$$
$$= \frac{\pm 1}{B_n(\alpha_{n+1}B_n + B_{n-1})},$$

on using (10). Since $\alpha_{n+1} > q_{n+1}$, we have

$$\left| \alpha - \frac{A_n}{B_n} \right| < \frac{1}{B_n B_{n+1}}. \tag{16}$$

The numbers B_0, B_1, B_2, ... are strictly increasing natural numbers; hence B_n increases indefinitely with n, and (16) proves that $\dfrac{A_n}{B_n}$ has the limit α as n increases indefinitely. This is the property which makes the word 'convergent' appropriate; $\dfrac{A_n}{B_n}$ converges to the value of the original number α as n increases indefinitely.

The representation of an irrational number by an infinite continued fraction suggests another question. In what precedes, the partial quotients q_0, q_1, q_2, ... were determined by the number α from which we started. Now suppose we select *any* infinite sequence of numbers q_0, q_1, q_2, ... , all of which are natural numbers except possibly the first, which may be any integer. Can we attach a meaning to the infinite continued fraction

$$q_0 + \frac{1}{q_1 +} \frac{1}{q_2 +} \cdots ?$$

If we can, will the resulting number be irrational, and will this continued fraction coincide with the one obtained by applying our former process to the number in question? Until we have settled these points, our theory is a very incomplete one.

In fact, the answers to these questions are as simple as one could wish. If one forms a continued fraction from *any* infinite sequence of natural numbers q_1, q_2, ... , preceded by any integer q_0, then the corresponding sequence of convergents has a limit. Perhaps the easiest proof is to consider the sequence formed by the even convergents $\dfrac{A_0}{B_0}$, $\dfrac{A_2}{B_2}$, ... This is an increasing sequence, and is bounded above, since all these are less than $\dfrac{A_1}{B_1}$ (for example). Hence, by the most fundamental of all propositions concerning limits, the sequence has a limit. Similarly the sequence formed by the odd convergents has a limit. Also the two limits are equal, since by (12) the difference between two consecutive convergents has the limit zero. Thus we can attach a meaning to *any* infinite continued fraction. If we denote the limit by α, then the continued fraction is in fact that which would arise from developing α in the way we considered originally at the beginning of this section. For the value of the infinite continued fraction

$$\frac{1}{q_1 +} \frac{1}{q_2 +} \cdots$$

is between 0 and 1; hence q_0 must be the integral part of α. If we write $\alpha = q_0 + \dfrac{1}{\alpha_1}$, we find that q_1 must be the integral part of α_1, and so on. In other words, the continued fraction is *unique*. In particular, the number defined by any infinite continued fraction must be irrational, for the continued fraction development of a rational number always terminates.

It now appears that infinite continued fractions provide not only *representations* for given irrational numbers, but a means of *constructing* irrational numbers. One way of describing the position is to say that the continued fraction process sets up a one-to-one correspondence between (i) all irrational numbers greater than 1, and (ii) all infinite sequences q_0, q_1, q_2, ... of natural numbers.

7. Diophantine approximation

The continued fraction process provides us with an infinite sequence of rational approximations to a given irrational number α, namely the convergents. Some information as to how rapidly they approach α is provided by the inequality (16). This implies, in particular, that if $\dfrac{x}{y}$ is any one of the convergents to α, then

$$\left| \alpha - \frac{x}{y} \right| < \frac{1}{y^2}. \tag{17}$$

We have here a simple result on Diophantine approximation: the branch of mathematics which is concerned with approximation to irrational numbers by means of rational numbers.

It is possible to prove, by rather more detailed arguments, that there are slightly better inequalities which are still satisfied by an infinity of rational approximations. In the first place, one can prove that of every two successive convergents, one at least satisfies

$$\left| \alpha - \frac{x}{y} \right| < \frac{1}{2y^2}.$$

Hence this inequality also is satisfied by an infinity of rational approximations. An inequality which is a little better still is satisfied by at least one out of every *three* successive convergents, namely

$$\left| \alpha - \frac{x}{y} \right| < \frac{1}{\sqrt{5}y^2}. \tag{18}$$

So any irrational number α has an infinity of rational approximations which satisfy (18), a result first proved by Hurwitz in 1891. Further than this one cannot go. There are irrational numbers for which any more precise inequality, say

$$\left| \alpha - \frac{x}{y} \right| < \frac{1}{ky^2}, \quad \text{where} \quad k > \sqrt{5}, \tag{19}$$

has only a finite number of solutions in integers x and y. The simplest example of such a number is the one given by the special continued fraction

$$\theta = 1 + \frac{1}{1+} \frac{1}{1+} \frac{1}{1+} \cdots .$$

This number has the property that any inequality of the form (19), with θ in place of α, has only a finite number of solutions. The actual value of θ is easily found from the fact that

$$\theta = 1 + \frac{1}{\theta}, \quad \text{or} \quad \theta^2 - \theta - 1 = 0.$$

Solving this quadratic equation, we obtain $\theta = \frac{1}{2}(1 + \sqrt{5})$, since the negative root is to be rejected.

The proofs of the various results which have just been mentioned are not especially difficult, but for them we must refer the reader to the literature cited in the Notes (p. 114).

8. *Quadratic irrationals*

The simplest and most familiar irrational numbers are the quadratic irrationals, that is, the numbers which arise as the solutions of quadratic equations with integral coefficients. In particular, the square root of any natural number N, not a perfect square, is a quadratic irrational, since it is a solution of the equation $x^2 - N = 0$. The continued fractions for quadratic irrationals have remarkable properties, which we shall now investigate.

Let us begin with a few numerical examples. Take first $\sqrt{2}$, as a very simple one. Since the integral part of $\sqrt{2}$ is 1, the first term q_0 of the continued fraction is 1, and the first step in the development consists in writing

$$\sqrt{2} = 1 + \frac{1}{\alpha_1}.$$

Here

$$\alpha_1 = \frac{1}{\sqrt{2}-1} = \sqrt{2} + 1.$$

The integral part of α_1 is 2, and so the next step is to write

$$\alpha_1 = 2 + \frac{1}{\alpha_2}.$$

Here

$$\alpha_2 = \frac{1}{\alpha_1 - 2} = \frac{1}{\sqrt{2} - 1} = \sqrt{2} + 1.$$

Since α_2 has turned out to be the same as α_1, there is no need of further calculation, for the subsequent steps will all be the same as the last step. All the subsequent terms of the continued fraction will be 2, and we have

$$\sqrt{2} = 1 + \frac{1}{2+} \frac{1}{2+} \frac{1}{2+} \cdots.$$

A few more examples are:

$$\sqrt{3} = 1 + \frac{1}{1+} \frac{1}{2+} \frac{1}{1+} \frac{1}{2+} \cdots,$$

$$\sqrt{5} = 2 + \frac{1}{4+} \frac{1}{4+} \frac{1}{4+} \cdots,$$

$$\sqrt{6} = 2 + \frac{1}{2+} \frac{1}{4+} \frac{1}{2+} \frac{1}{4+} \cdots.$$

To take a slightly more complicated example, consider the number

$$\alpha = \frac{24 - \sqrt{15}}{17}.$$

Since $\sqrt{15}$ lies between 3 and 4, the integral part of α is 1. The first step is to write

$$\alpha = 1 + \frac{1}{\alpha_1}.$$

Here

$$\alpha_1 = \frac{1}{\alpha - 1} = \frac{17}{7 - \sqrt{15}} = \frac{7 + \sqrt{15}}{2}.$$

The integral part of α_1 is 5, so

$$\alpha_1 = 5 + \frac{1}{\alpha_2},$$

where

$$\alpha_2 = \frac{1}{\alpha_1 - 5} = \frac{2}{\sqrt{15} - 3} = \frac{\sqrt{15} + 3}{3}.$$

The integral part of α_2 is 2, so

$$\alpha_2 = 2 + \frac{1}{\alpha_3},$$

where

$$\alpha_3 = \frac{1}{\alpha_2 - 2} = \frac{3}{\sqrt{15} - 3} = \frac{\sqrt{15} + 3}{2}.$$

The integral part of α_3 is 3, so

$$\alpha_3 = 3 + \frac{1}{\alpha_4},$$

where

$$\alpha_4 = \frac{1}{\alpha_3 - 3} = \frac{2}{\sqrt{15} - 3} = \frac{\sqrt{15} + 3}{3}.$$

Since $\alpha_4 = \alpha_2$, the last two steps will be repeated over and over again, and the continued fraction is

$$\frac{24 - \sqrt{15}}{17} = 1 + \frac{1}{5+} \frac{1}{2+} \frac{1}{3+} \frac{1}{2+} \frac{1}{3+} \cdots.$$

We can abbreviate this to

$$1, 5, \overline{2, 3},$$

where the bar indicates the period, which is repeated indefinitely. With this short notation, the previous examples take the form:

$$\sqrt{2} = 1, \overline{2}; \quad \sqrt{3} = 1, \overline{1, 2}; \quad \sqrt{5} = 2, \overline{4}; \quad \sqrt{6} = 2, \overline{2, 4}.$$

In each of these cases, it is found that a complete quotient α_n is reached which is the same as some previous complete quotient α_m. From that point onwards, the continued fraction is *periodic*. The terms consist of the numbers from q_m to q_{n-1}, repeated over and over again. The general theorem that *any quadratic irrational number has a continued fraction which is periodic after a certain stage* was first proved by Lagrange in 1770, though the fact was known to earlier mathematicians. We shall prove this theorem in §10, after first considering purely periodic continued fractions in §9.

A table of the continued fractions for $\sqrt{N}$, for $N=2, 3, \ldots , 50$ (excluding perfect squares) is given on p. 105. For simplicity the bar is omitted from the period, which consists of all the numbers after the first term. It will be seen that all these continued fractions have certain features in common, and the reason for this will become plain in the course of the next section.

For purposes of numerical calculation, the process which we used in the above examples can be simplified by restricting one's attention to the *integers* involved, and arranging the work in a more concise form.

9. *Purely periodic continued fractions*

It so happens that in each of the numerical examples considered above, the continued fraction is not periodic from the beginning, but only after a certain stage. But we can easily give examples of purely periodic continued fractions; for example, if we add 1 to the continued fraction for $\sqrt{2}$, we obtain

$$\sqrt{2} + 1 = 2 + \frac{1}{2+} \frac{1}{2+} \cdots ,$$

which is purely periodic. Similarly

$$\sqrt{6} + 2 = 4 + \frac{1}{2+} \frac{1}{4+} \frac{1}{2+} \cdots .$$

The numbers represented by purely periodic continued fractions are a particular kind of quadratic irrational, and we shall now investigate how these numbers can be characterized.

Let us begin with a particular example. Consider some purely periodic continued fraction, say

$$\alpha = 4 + \frac{1}{1+} \frac{1}{3+} \frac{1}{4+} \frac{1}{1+} \frac{1}{3+} \cdots .$$

This definition of α can also be written in the form

$$\alpha = 4 + \frac{1}{1+} \frac{1}{3+} \frac{1}{\alpha}. \tag{20}$$

We have here an equation for α, which, when worked out, will in fact be a quadratic equation. To see what this equation is, compare the above relation with (13), of which it is a special case, with $\alpha_{n+1}=\alpha$. It follows from the general formula (14) that

$$\alpha = \frac{19\alpha+5}{4\alpha+1}, \tag{21}$$

because $\frac{19}{4}$ and $\frac{5}{1}$ are the two convergents preceding the term $\frac{1}{\alpha}$ in (20). Thus the quadratic equation satisfied by α is

$$4\alpha^2-18\alpha-5=0. \tag{22}$$

It will be instructive to consider, at the same time as α, the number β defined in the same way but with the period reversed, that is

$$\beta = 3 + \frac{1}{1+} \frac{1}{4+} \frac{1}{3+} \frac{1}{1+} \frac{1}{4+} \cdots.$$

The relation analogous to (20) is

$$\beta = 3 + \frac{1}{1+} \frac{1}{4+} \frac{1}{\beta}.$$

When we apply the general formula (14), we obtain

$$\beta = \frac{19\beta+4}{5\beta+1}, \tag{23}$$

since the two convergents are now $\frac{19}{5}$ and $\frac{4}{1}$. Hence the quadratic equation satisfied by the number β is

$$5\beta^2-18\beta-4=0. \tag{24}$$

This is obviously closely related to the previous equation (22) satisfied by α. Indeed, if we put $-\frac{1}{\beta}=\alpha$, the equation (24) is transformed into the equation (22). Hence the number $-\frac{1}{\beta}$ is one of the two roots of the quadratic equation (22). It cannot be the number α itself, because α and β are positive, and $-\frac{1}{\beta}$ is negative. Hence $-\frac{1}{\beta}$ is the *second* root of the equation (22). This second root is

called the *algebraic conjugate* of α, or simply the conjugate of α.
Denoting the conjugate of α by α', we have $\alpha' = -\dfrac{1}{\beta}$.

The above argument is really quite general. In the case of any purely periodic continued fraction, say

$$\alpha = q_0 + \frac{1}{q_1+} \cdots \frac{1}{q_n+} \frac{1}{\alpha},$$

the equation corresponding to (21) is

$$\alpha = \frac{A_n\alpha + A_{n-1}}{B_n\alpha + B_{n-1}}.$$

If the number β is then defined by reversing the period, the equation corresponding to (23) is

$$\beta = \frac{A_n\beta + B_n}{A_{n-1}\beta + B_{n-1}},$$

this being a consequence of the fact that the value of $[q_0, \ldots, q_n]$ is unchanged if the terms are taken in the opposite order (§3). The two quadratic equations for α and β are related in just the same way as above, and $-\dfrac{1}{\beta}$ is the conjugate of α. Since β is greater than 1, the number $-\dfrac{1}{\beta}$ lies between -1 and 0. Hence *any purely periodic continued fraction represents a quadratic irrational number α which is greater than 1, and whose conjugate lies between -1 and 0. This conjugate is $-\dfrac{1}{\beta}$, where β is defined by the continued fraction with the reversed period.*

It is a remarkable fact that this simple property completely characterizes the numbers represented by purely periodic continued fractions; as we shall now prove, any quadratic irrational number which satisfies the condition does have a purely periodic continued fraction. This seems to have first been proved explicitly by Galois in 1828, though the result was implicit in the earlier work of Lagrange.

We shall call a quadratic irrational number α *reduced* if $\alpha > 1$ and if the conjugate of α, denoted by α', satisfies $-1 < \alpha' < 0$. Our object is to prove that the continued fraction for α is purely periodic. Naturally the proof is more difficult than that of the result proved above, where we began with the continued fraction; moreover, the proof is not of such a nature that it can be adequately illustrated by an example.

We begin by investigating the form of a reduced quadratic irrational number. We know that α satisfies some quadratic equation

$$a\alpha^2 + b\alpha + c = 0,$$

where a, b, c are integers. Solving this equation, we can express α in the form

$$\alpha = \frac{-b \pm \sqrt{b^2 - 4ac}}{2a} = \frac{P \pm \sqrt{D}}{Q},$$

where P and Q are integers, and D is a positive integer which is not a perfect square. We can suppose that the $+$ sign is attached to $\sqrt{D}$, for if it were the $-$ sign, we could change it to the $+$ sign by changing the signs of both the numbers P and Q. So

$$\alpha = \frac{P + \sqrt{D}}{Q}, \tag{25}$$

and the conjugate α' of α, being the other root of the quadratic equation, is given by

$$\alpha' = \frac{P - \sqrt{D}}{Q}.$$

We note that

$$\frac{P^2 - D}{Q} = \frac{b^2 - (b^2 - 4ac)}{2a} = 2c,$$

so that $P^2 - D$ is a multiple of Q.

Since α is supposed to be *reduced*, we have $\alpha > 1$ and $-1 < \alpha' < 0$. This implies that

(i) $\alpha - \alpha' > 0$, that is $\dfrac{\sqrt{D}}{Q} > 0$, whence $Q > 0$;

(ii) $\alpha + \alpha' > 0$, that is $\dfrac{P}{Q} > 0$, whence $P > 0$;

(iii) $\alpha' < 0$, that is $P < \sqrt{D}$;

(iv) $\alpha > 1$, that is $Q < P + \sqrt{D} < 2\sqrt{D}$.

Thus a reduced quadratic irrational number α is of the form (25), where P and Q are natural numbers satisfying*

$$P<\sqrt{D}, \quad Q<2\sqrt{D}, \tag{26}$$

and also satisfying the condition that P^2-D is a multiple of Q.

Now let α be developed into a continued fraction. The first step in the process of development is to express α in the form

$$\alpha = q_0 + \frac{1}{\alpha_1}, \tag{27}$$

where q_0 is the integral part of α, and $\alpha_1 > 1$. It is easy to see that α_1 is again a reduced quadratic irrational, for the equation (27) implies that the conjugates of α and α_1 are connected by the similar relation

$$\alpha' = q_0 + \frac{1}{\alpha_1'}.$$

So

$$\alpha_1' = -\frac{1}{q_0-\alpha'},$$

and since α' is negative, and q_0 is a natural number, we have $q_0-\alpha'>1$, and therefore α_1' lies between -1 and 0. Similarly, all the subsequent complete quotients $\alpha_2, \alpha_3, \ldots$ in the development are reduced quadratic irrationals.

As regards the form of α_1 we have

$$\frac{1}{\alpha_1} = \alpha-q_0 = \frac{P+\sqrt{D}}{Q} - q_0 = \frac{P-Qq_0 + \sqrt{D}}{Q}.$$

Let $P_1 = -P+Qq_0$. Then

$$\alpha_1 = \frac{Q}{-P_1+\sqrt{D}} = \frac{P_1+\sqrt{D}}{Q_1},$$

where Q_1 is defined by

$$D-P_1{}^2 = QQ_1. \tag{28}$$

Note that Q_1 is an integer, since P^2-D is a multiple of Q and $P_1 \equiv -P \pmod{Q}$. We have

$$\alpha_1 = \frac{P_1+\sqrt{D}}{Q_1}, \tag{29}$$

*It must not be supposed that every number α satisfying these conditions is reduced, for these conditions do not necessarily ensure that $\alpha' > -1$.

and since α_1 is reduced, the integers P_1 and Q_1 are positive, and satisfy the conditions (26). Moreover, $P_1{}^2 - D$ is a multiple of Q_1, by (28).

We are now in a position to see how the continued fraction process goes on. At the next step we start from α_1 instead of from α, but the process is just the same. Generally, each complete quotient has the form

$$\alpha_n = \frac{P_n + \sqrt{D}}{Q_n},$$

where P_n and Q_n are natural numbers which satisfy (26), and have the property that $P_n{}^2 - D$ is a multiple of Q_n. There are only a finite number of possibilities for P_n and Q_n by (26), and eventually we must come to some pair of values which has occurred before. That is, we must come to some complete quotient which is the same as some earlier one, and from this point onwards the continued fraction is periodic.

We have still to prove that the continued fraction is *purely* periodic, that is, periodic from the beginning. To prove this, we shall show that if $\alpha_n = \alpha_m$, then $\alpha_{n-1} = \alpha_{m-1}$, and in this way we shall be able to work backwards to the beginning of the continued fraction. The proof depends on the fact that it is possible to relate the partial quotients q_n not only to the complete quotients α_n but also, in a somewhat similar way, to their conjugates. The relation between any complete quotient and the next is

$$\alpha_n = q_n + \frac{1}{\alpha_{n+1}}.$$

The same relation must connect their conjugates, so that

$$\alpha_n{}' = q_n + \frac{1}{\alpha_{n+1}{}'}.$$

Since each conjugate lies between -1 and 0, let us introduce the symbol β_n for $-\dfrac{1}{\alpha_n{}'}$. Then each of the numbers β_n is greater than 1. The last relation takes the form

$$-\frac{1}{\beta_n} = q_n - \beta_{n+1}, \quad \text{or} \quad \beta_{n+1} = q_n + \frac{1}{\beta_n}.$$

It now follows from the last relation that q_n, in addition to being the integral part of α_n, can also be interpreted as being the integral part of β_{n+1}.

Now suppose that α_n and α_m are two equal complete quotients, where $m<n$. Then their conjugates $\alpha_n{}'$ and $\alpha_m{}'$ are also equal, and therefore $\beta_n=\beta_m$. By the result just proved, q_{n-1} is the integral part of β_n, and q_{m-1} is the integral part of β_m. Hence $q_{n-1}=q_{m-1}$. But

$$\alpha_{n-1} = q_{n-1} + \frac{1}{\alpha_n}, \quad \alpha_{m-1} = q_{m-1} + \frac{1}{\alpha_m}.$$

Hence $\alpha_{n-1}=\alpha_{m-1}$. Repeating the argument, we obtain $\alpha_{n-2}=\alpha_{m-2}$, and so on until we reach the fact that α_{n-m} is the same as α itself. Putting $n-m=r$, we have

$$\alpha = q_0 + \frac{1}{q_1+} \cdots \frac{1}{q_{r-1}+} \frac{1}{\alpha},$$

and this shows that the continued fraction for α is purely periodic. We have proved the result which is the main object of this section, namely that the purely periodic continued fractions represent precisely the reduced quadratic irrationals.

It is now possible to see why the continued fractions for $\sqrt{N}$, where N is a natural number, not a perfect square, are all of the special type which we see in the table. The continued fraction for $\sqrt{N}$ certainly cannot be purely periodic, because the conjugate of $\sqrt{N}$ is $-\sqrt{N}$, and this does not lie between -1 and 0. But consider the number $\sqrt{N}+q_0$, where q_0 is the integral part of $\sqrt{N}$. The conjugate of this number is $-\sqrt{N}+q_0$, which does lie between -1 and 0. Hence the continued fraction for $\sqrt{N}+q_0$ is purely periodic, and since it obviously begins with $2q_0$, it is of the form

$$\sqrt{N} + q_0 = 2q_0 + \frac{1}{q_1+} \cdots \frac{1}{q_n+} \frac{1}{2q_0+} \cdots . \tag{30}$$

According to the result proved earlier in this section, the continued fraction formed with the period reversed, that is

$$q_n + \frac{1}{q_{n-1}+} \cdots \frac{1}{q_1+} \quad \frac{1}{2q_0+} \quad \frac{1}{q_n+} \cdots ,$$

must represent $-\dfrac{1}{\alpha'}$, where $\alpha=\sqrt{N}+q_0$. Now $\alpha'=-\sqrt{N}+q_0$, hence

$$-\frac{1}{\alpha'} = \frac{1}{\sqrt{N-q_0}} = q_1 + \frac{1}{q_2+} \cdots \frac{1}{q_n+} \frac{1}{2q_0+} \cdots,$$

by (30). Comparing the last two continued fractions (and recalling the fact that the development of a number is unique), we see that

$$q_n = q_1, \quad q_{n-1} = q_2, \quad \cdots.$$

Hence *the continued fraction for $\sqrt{N}$ is necessarily of the form*

$$q_0, \overline{q_1, q_2, \cdots, q_2, q_1, 2q_0}.$$

The period begins immediately after the first term q_0, and it consists of a symmetrical part $q_1, q_2, \cdots, q_2, q_1$, followed by the number $2q_0$. The symmetrical part may or may not have a central term; for example, in

$$\sqrt{54} = 7, \overline{2, 1, 6, 1, 2, 14}$$

there is a central term, whereas in

$$\sqrt{53} = 7, \overline{3, 1, 1, 3, 14}$$

there is none. The symmetrical part of the period may of course be absent, in which case the period reduces to the single number $2q_0$, as in $\sqrt{2} = 1, \overline{2}$.

10. *Lagrange's theorem*

We can now prove the general theorem of Lagrange that *any* quadratic irrational has a continued fraction which is periodic from some point onwards. It will be enough to prove that when *any* quadratic irrational α is developed into a continued fraction, we reach sooner or later a complete quotient α_n which is a *reduced* quadratic irrational; for then the continued fraction will be periodic from that point onwards.

The relation between α itself and one of the complete quotients is given by the familiar formula (14):

$$\alpha = \frac{\alpha_{n+1}A_n + A_{n-1}}{\alpha_{n+1}B_n + B_{n-1}}.$$

Since α and α_{n+1} are quadratic irrationals, and A_n, B_n, A_{n-1}, B_{n-1} are integers (indeed, natural numbers), the same relation must hold between α' and α_{n+1}'. Solving it to express α_{n+1}' in terms of α', we obtain

	TABLE I			
N	Continued fraction for $\sqrt{N}$	x	y	$x^2 - Ny^2$
2	1; 2	1	1	-1
3	1; 1, 2	2	1	$+1$
5	2; 4	2	1	-1
6	2; 2, 4	5	2	$+1$
7	2; 1, 1, 1, 4	8	3	$+1$
8	2; 1, 4	3	1	$+1$
10	3; 6	3	1	-1
11	3; 3, 6	10	3	$+1$
12	3; 2, 6	7	2	$+1$
13	3; 1, 1, 1, 1, 6	18	5	-1
14	3; 1, 2, 1, 6	15	4	$+1$
15	3; 1, 6	4	1	$+1$
17	4; 8	4	1	-1
18	4; 4, 8	17	4	$+1$
19	4; 2, 1, 3, 1, 2, 8	170	39	$+1$
20	4; 2, 8	9	2	$+1$
21	4; 1, 1, 2, 1, 1, 8	55	12	$+1$
22	4; 1, 2, 4, 2, 1, 8	197	42	$+1$
23	4; 1, 3, 1, 8	24	5	$+1$
24	4; 1, 8	5	1	$+1$
26	5; 10	5	1	-1
27	5; 5, 10	26	5	$+1$
28	5; 3, 2, 3, 10	127	24	$+1$
29	5; 2, 1, 1, 2, 10	70	13	-1
30	5; 2, 10	11	2	$+1$
31	5; 1, 1, 3, 5, 3, 1, 1, 10	1520	273	$+1$
32	5; 1, 1, 1, 10	17	3	$+1$
33	5; 1, 2, 1, 10	23	4	$+1$
34	5; 1, 4, 1, 10	35	6	$+1$
35	5; 1, 10	6	1	$+1$
37	6; 12	6	1	-1
38	6; 6, 12	37	6	$+1$
39	6; 4, 12	25	4	$+1$
40	6; 3, 12	19	3	$+1$
41	6; 2, 2, 12	32	5	-1
42	6; 2, 12	13	2	$+1$
43	6; 1, 1, 3, 1, 5, 1, 3, 1, 1, 12	3482	531	$+1$
44	6; 1, 1, 1, 2, 1, 1, 1, 12	199	30	$+1$
45	6; 1, 2, 2, 2, 1, 12	161	24	$+1$
46	6; 1, 3, 1, 1, 2, 6, 2, 1, 1, 3, 1, 12	24335	3588	$+1$
47	6; 1, 5, 1, 12	48	7	$+1$
48	6; 1, 12	7	1	$+1$
50	7; 14	7	1	-1

$$\alpha_{n+1}' = -\frac{B_{n-1}\alpha' - A_{n-1}}{B_n\alpha' - A_n} = -\frac{B_{n-1}}{B_n}\left(\frac{\alpha' - A_{n-1}/B_{n-1}}{\alpha' - A_n/B_n}\right).$$

What does this tell us about the magnitude of α_{n+1}' when n is large? Both $\dfrac{A_n}{B_n}$ and $\dfrac{A_{n-1}}{B_{n-1}}$ tend to the limit α as n increases indefinitely, and consequently the fraction in brackets has the limit 1. Also B_{n-1} and B_n are positive, and so α_{n+1}' is ultimately negative. Further, the numbers $\dfrac{A_n}{B_n}$ are alternately less than α and greater than α (§4), and therefore the fraction in brackets is alternately slightly less than 1 and slightly greater than 1. If we select a value of n for which it is slightly less than 1, and note also that $B_{n-1} < B_n$, we see that α_{n+1}' lies between -1 and 0. For this value of n, the number α_{n+1} is a reduced quadratic irrational. Consequently the continued fraction will be purely periodic from that stage onwards (or possibly from some earlier stage). This establishes Lagrange's theorem.

There are not many irrational numbers, other than quadratic irrationals, whose continued fractions are known to have any features of regularity. One such number is $\dfrac{e-1}{e+1}$, where e is the basis of the natural logarithms: $e = 2\cdot71828\ldots$. The continued fraction is

$$\frac{e-1}{e+1} = \frac{1}{2+}\ \frac{1}{6+}\ \frac{1}{10+}\ \frac{1}{14+}\cdots,$$

the terms forming an arithmetical progression. More generally, if k is any positive integer,

$$\frac{e^{2/k} - 1}{e^{2/k} + 1} = \frac{1}{k+}\ \frac{1}{3k+}\ \frac{1}{5k+}\ \frac{1}{7k+}\cdots.$$

These results were found by Euler in 1737. The continued fraction for e itself is a little more complicated:

$$e = 2 + \frac{1}{1+}\ \frac{1}{2+}\ \frac{1}{1+}\ \frac{1}{1+}\ \frac{1}{4+}\ \frac{1}{1+}\ \frac{1}{1+}\ \frac{1}{6+}\cdots,$$

where the numbers 2, 4, 6, ... are separated by two 1's each time. This also was found by Euler.

Very little is known about the continued fractions for algebraic numbers, apart from quadratic irrationals. We do not know, for example, whether the terms in the continued fraction for $\sqrt[3]{2}$, which begins

$$\sqrt[3]{2} = 1 + \frac{1}{3+} \frac{1}{1+} \frac{1}{5+} \frac{1}{1+} \frac{1}{1+} \frac{1}{4+} \frac{1}{1+} \cdots ,$$

are bounded or not; and there seems to be no method by which such a problem can be attacked. Some results are known about Diophantine approximation to algebraic numbers (see VII.6), and these imply that the terms of the continued fractions for such numbers cannot increase with more than a certain degree of rapidity. But the results found in this way are probably far from the real truth.

11. *Pell's equation*

This is the equation

$$x^2 - Ny^2 = 1, \quad \text{or} \quad x^2 = Ny^2 + 1, \tag{31}$$

where N is a natural number which is not a perfect square. (The equation is of no interest when N is a perfect square, since the difference of two perfect squares can never be 1, except in the case $1^2 - 0^2$.) It is a remarkable fact that Pell's equation always has a solution in natural numbers x and y, and indeed has infinitely many such solutions.

References to individual cases of Pell's equation occur scattered throughout the history of mathematics. The most curious of these occurrences is in the so-called Cattle Problem of Archimedes, published by Lessing in 1773 from a manuscript in the library of Wolfenbüttel. The problem is stated to have been propounded by Archimedes to Eratosthenes, and most of the experts who have investigated the matter have reached the conclusion that the problem was in fact invented by Archimedes. It contains eight unknowns (numbers of cattle of various kinds) which satisfy seven linear equations, together with two conditions which assert that certain numbers are perfect squares. After some elementary algebra, the problem reduces to that of solving the equation

$$t^2 - 4,729,494 \, u^2 = 1,$$

the least solution of which (given by Amthor in 1880) is a number u of forty-one digits. The least solution of the original problem, deduced from this, consists of numbers with hundreds of thousands of digits. There is no evidence that the ancients could solve the problem, but the mere fact that they propounded it suggests that they may well have had some knowledge about Pell's equation which has not survived.

In modern times, the first systematic method for solving Pell's equation was given by Lord Brouncker* in 1657. It is essentially that of developing $\sqrt{N}$ into a continued fraction, as explained below. About the same time, Frénicle de Bessy (in a work which has not survived) tabulated solutions of (31) for all values of N up to 150, and challenged Brouncker to solve the equation $x^2 - 313y^2 = 1$. Brouncker, in reply, gave a solution (in which x has sixteen digits), which he said he had found by his method within an hour or two. Both Wallis, when expounding Brouncker's method, and Fermat, in commenting on Wallis's work, claimed to have proved that the equation is always soluble. Fermat seems to have been the first to state categorically that there are infinitely many solutions. The first published proof was that of Lagrange, which appeared in about 1766. The name of Pell was attached to the equation by Euler under a misapprehension; he thought that the method of solution given by Wallis was due to John Pell, another English mathematician of the same period.

A solution of Pell's equation is easily obtained in terms of the continued fraction for $\sqrt{N}$. We saw in §9 that this is of the form

$$\sqrt{N} = q_0 + \frac{1}{q_1+} \cdots \frac{1}{q_n+} \frac{1}{2q_0+} \frac{1}{q_1+} \cdots .$$

(We saw also that $q_n = q_1$, etc., but this is of no importance at the moment.) Now let $\dfrac{A_{n-1}}{B_{n-1}}$ and $\dfrac{A_n}{B_n}$ be the two convergents coming immediately before the term $2q_0$, that is

$$\frac{A_{n-1}}{B_{n-1}} = q_0 + \frac{1}{q_1+} \cdots \frac{1}{q_{n-1}}, \quad \frac{A_n}{B_n} = q_0 + \frac{1}{q_1+} \cdots \frac{1}{q_n}.$$

By the formula (14), we have

$$\sqrt{N} = \frac{\alpha_{n+1} A_n + A_{n-1}}{\alpha_{n+1} B_n + B_{n-1}},$$

*William Brouncker (1620?-84) succeeded his father as second Viscount Brouncker, of Castle Lyons in Ireland, in 1667. Readers of the *Diary* will recall that Pepys had a low opinion of his moral character. But his mathematical achievements are very creditable.

where α_{n+1} is the complete quotient after q_n, that is,

$$\alpha_{n+1} = 2q_0 + \cfrac{1}{q_1+} \ldots = \sqrt{N} + q_0.$$

Substituting this value for α_{n+1}, and multiplying up, we obtain

$$\sqrt{N}\,(\sqrt{N} + q_0)\,B_n + \sqrt{N}\,B_{n-1} = (\sqrt{N} + q_0)\,A_n + A_{n-1}.$$

Since $\sqrt{N}$ is irrational, and all the other numbers are integers, this equation implies the two equations

$$NB_n = q_0\,A_n + A_{n-1},$$
$$q_0\,B_n + B_{n-1} = A_n.$$

These may be regarded as expressing A_{n-1} and B_{n-1} in terms of A_n and B_n:

$$A_{n-1} = NB_n - q_0\,A_n, \quad B_{n-1} = A_n - q_0\,B_n.$$

Now substitute in (10). We obtain

$$A_n\,(A_n - q_0\,B_n) - B_n\,(NB_n - q_0\,A_n) = (-1)^{n-1},$$

or

$$A_n{}^2 - NB_n{}^2 = (-1)^{n-1}. \tag{32}$$

Hence $x = A_n$ and $y = B_n$ provides a solution of the equation

$$x^2 - Ny^2 = (-1)^{n-1}.$$

If n is odd, we have a solution of Pell's equation. If not, we observe that the same argument would apply to the two convergents at the end of the next period. Since the term q_n, where it occurs for the second time, would be q_{2n+1} if the terms were numbered consecutively, we have to change n in (32) into $2n+1$, giving

$$A_{2n+1}{}^2 - NB_{2n+1}{}^2 = (-1)^{2n} = 1.$$

So in any case the equation (31) is soluble in natural numbers x and y.

We illustrate the theory by two numerical examples, one for which n is odd and one for which n is even. Take first $N=21$. The continued fraction (see Table I, p. 105) is

$$\sqrt{21} = 4, \overline{1, 1, 2, 1, 1, 8},$$

and $n = 5$. The convergents are

$$\tfrac{4}{1}, \tfrac{5}{1}, \tfrac{9}{2}, \tfrac{23}{5}, \tfrac{32}{7}, \tfrac{55}{12}, \ldots ,$$

and $x = 55$, $y = 12$ gives a solution of

$$x^2 - 21y^2 = 1.$$

Take next $N = 29$. The continued fraction is

$$\sqrt{29} = 5, \overline{2, 1, 1, 2, 10},$$

and $n = 4$. The convergents are

$$\tfrac{5}{1}, \tfrac{11}{2}, \tfrac{16}{3}, \tfrac{27}{5}, \tfrac{70}{13}, \cdots,$$

and $x = 70$, $y = 13$ gives a solution of

$$x^2 - 29y^2 = -1.$$

To obtain a solution of the equation with 1, and not -1, we continue the series of convergents until we reach $\dfrac{A_9}{B_9}$ (since $2n+1 = 9$).

Now $\dfrac{A_4}{B_4} = \dfrac{70}{13}$, and the next few convergents are

$$\tfrac{727}{135}, \tfrac{1524}{283}, \tfrac{2251}{418}, \tfrac{3775}{701}, \tfrac{9801}{1820}.$$

Hence $x = 9801$, $y = 1820$ gives a solution of

$$x^2 - 29y^2 = 1.$$

It can be proved that the process which has been explained above always gives the *smallest* solution of Pell's equation. The smallest solutions of $x^2 - Ny^2 = \pm 1$ are given in Table 1 up to $N = 50$.

There are several other facts about Pell's equation which can be proved by the methods we have used in this section. The first is that the equation has infinitely many solutions, and that these are given by all the convergents which correspond to the terms q_n at the end of each period. If n is odd, that is, if the continued fraction has a central term (as in the example with $\sqrt{21}$) all these are solutions of the equation with $+1$. If n is even, that is if there is no central term (as in the example with $\sqrt{29}$), the convergents just specified give alternately solutions with -1 and $+1$.

The later solutions can also be obtained from the first solution by direct calculation, without developing further the continued fraction. If x_0, y_0 is the smallest solution of $x^2 - Ny^2 = \pm 1$, given by the convergent $\dfrac{A_n}{B_n}$, then the general solution x, y is given by

$$x+y\sqrt{N}=(x_0+y_0\sqrt{N})^r,$$

where $r=1, 2, 3, \ldots$. Thus, in the example with $\sqrt{29}$ it will be found that

$$9801+1820\sqrt{29}=(70+13\sqrt{29})^2.$$

The distinction between the cases when n is odd or even raises problems to which no complete answer is known. No way of completely characterizing the numbers N for which n is even has been found. If the equation $x^2-Ny^2=-1$ is soluble, the congruence

$$x^2+1\equiv0 \;(\mathrm{mod}\; N)$$

is soluble. It follows that N cannot be divisible by 4 and also cannot be divisible by any prime of the form $4k+3$ (III.3). In fact, as we shall see later (VI.5), N is representable as u^2+v^2, where u and v are relatively prime. This, then, is a necessary condition for the solubility of $x^2-Ny^2=-1$, but it is not sufficient; for example the number $N=34$ satisfies the condition, but the equation $x^2-34y^2=-1$ is insoluble.

The solutions of the more general equation

$$x^2-Ny^2=\pm M,$$

where M is a positive integer less than $\sqrt{N}$, are also closely related to the continued fraction for $\sqrt{N}$. It can be proved that *every solution of every such equation comes from some convergent in the continued fraction for $\sqrt{N}$.*

12. *A geometrical interpretation of continued fractions*

A striking geometrical interpretation of the continued fraction for an irrational number was given by Klein in 1895. Suppose α is an irrational number, which we suppose for simplicity to be positive. Consider all points in the plane whose coordinates are positive integers, and imagine that pegs are inserted in the plane at all such points. The line $y=\alpha x$ does not pass through any of them. Imagine a string drawn along the line, with one end fixed at an infinitely remote point on the line. If the other end of the string, at the origin, is pulled away from the line on one side, the string will catch on certain pegs: if it is pulled away from the line on the other side, the string will catch on certain other pegs. One set of pegs (those below the line) consists of the points with co-ordinates

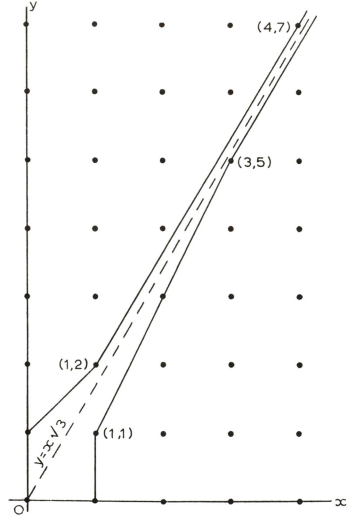

Fig. 3

(B_0, A_0), (B_2, A_2), ... , corresponding to the convergents which are less than α. The other set of pegs (those above the line) consists of the points with coordinates (B_1, A_1), (B_3, A_3), ... , corresponding to the convergents which are greater than α. Each of the two positions of the string forms a polygonal line, approaching the line $y = \alpha x$.

Figure 3 illustrates the case

$$\alpha = \sqrt{3} = 1 + \cfrac{1}{1+} \cfrac{1}{2+} \cfrac{1}{1+} \cfrac{1}{2+} \cdots .$$

Here the convergents are

$$\tfrac{1}{1}, \tfrac{2}{1}, \tfrac{5}{3}, \tfrac{7}{4}, \tfrac{19}{11}, \tfrac{26}{15}, \cdots .$$

The pegs below the line are at the points

$$(1, 1), (3, 5), (11, 19), \ldots ,$$

and the pegs above the line are at the points

$$(1, 2), (4, 7), (15, 26), \ldots .$$

Most of the elementary theorems about continued fractions have simple geometrical interpretations. If P_n denotes generally the point (B_n, A_n), the recurrence relations (8) and (9) state that the vector from P_{n-2} to P_n (two consecutive vertices on one of the polygonal lines) is an integral multiple of the vector from the origin O to P_{n-1}. The relation (10) can be interpreted as stating that the area of the triangle $OP_{n-1}P_n$ is always $\tfrac{1}{2}$. This can be deduced directly from the above construction with a string; for it is obvious that there cannot be any point with integral coordinates in the triangle $OP_{n-1}P_n$ other than the vertices themselves, and it is easy to prove that any triangle with this property has area $\tfrac{1}{2}$.

NOTES

The best account of continued fractions available in English is that in Chrystal's *Algebra*, vol. II, chs. 32–4. The standard work on the subject is Perron's *Die Lehre von den Kettenbrüchen* (Teubner, 1929). Proofs of the various results which are stated without proof in this chapter will be found in either Chrystal or Perron. On Diophantine approximation, the reader may consult Perron's *Irrationalzahlen* (Göschens Lehr-bücherei, vol. 1, 1947) or Niven's *Irrational Numbers* (Carus Math. Monographs no. 11, 1956) or Cassels's *Introduction to Diophantine Approximation* (Cambridge Math. Tracts no. 45, 1957).

§§1–6. Practically the whole of this theory is due to Euler.

§7. See Hardy and Wright, ch. 11, or Perron, §14.

§8. References to tables will be found in Perron, p. 100, or Dickson's *History*, vol. II, ch. 12. For abbreviated methods of calculating the continued fractions of quadratic irrationals, see Dickson's *History*, vol. II, p. 372.

§10. For proofs of the continued fractions for *e*, etc., see Perron §§31 and 64, or a note by C. S. Davis in *J. London Math. Soc.*, 20 (1945), 194–8.

§11. For the cattle problem, see Sir Thomas Heath, *Diophantus of Alexandria* (Cambridge, 1910), pp. 121–4, and Dickson's *History*, vol. II, pp. 342–5.

§12. See Klein's *Ausgewählte Kapitel der Zahlentheorie* (Teubner, 1907) pp. 17–25. The idea seems to be due to H. J. S. Smith (see his *Collected Math. Papers*, vol. 2, 146–7).

V

SUMS OF SQUARES

1. *Numbers representable by two squares*

The question what numbers are representable as the sum of two squares is a very old one; there are some statements bearing on it in the *Arithmetic* of Diophantus (about 250 A.D.), but their precise meaning is not clear. The true answer to the question was first given by the Dutch mathematician Albert Girard in 1625, and again by Fermat a little later. It is probable that Fermat had proofs of his results, but the first proofs we know of are those published by Euler in 1749.

It is an easy matter to rule out certain numbers as incapable of being represented as the sum of two squares. In the first place, the square of any even number is congruent to 0 (mod 4), and the square of any odd number is congruent to 1 (mod 4). Hence the sum of any two squares must be congruent either to $0+0$ or $0+1$ or $1+1$ (mod 4), that is either to 0 or 1 or 2 (mod 4). Thus any number which is of the form $4k+3$ cannot be the sum of two squares.

But we can go further than this. If a number N has a prime factor q which is of the form $4k+3$, the equation $x^2+y^2=N$ would imply the congruence $x^2 \equiv -y^2$ (mod q), and since -1 is a quadratic non-residue to the modulus q, this congruence holds only when $x \equiv 0$ and $y \equiv 0$ (mod q). Hence x and y are divisible by q, and N is divisible by q^2, and the equation $x^2+y^2=N$ can be divided throughout by q^2. If $N=q^2N_1$ and N_1 is still divisible by q then by the same argument it must be divisible by q^2, and so on, until eventually we find that the exact power of q which divides N must be even. Thus a number which is expressible as the sum of two squares must, when factorized into powers of primes, contain only even powers of primes of the form $4k+3$. This condition includes and supersedes the previous condition that N must not itself be of the form $4k+3$, for a number of the form $4k+3$ must contain some prime factor of that form to an odd power.

If we rule out the numbers which because of the condition just found cannot be sums of two squares, the remaining numbers begin:

$$1, \quad 2, \quad 4, \quad 5, \quad 8, \quad 9, \quad 10, \quad 13, \quad 16, \quad 17, \quad 18, \quad 20, \quad \ldots,$$

and the reader will find on trial that each of these *is* representable as the sum of two integral squares. This is true generally, and the criterion for representability of a number is that *any prime factor of N which is of the form $4k+3$ must divide N to an even power exactly*.

Our object now is to prove this result. An important part in the proof is played by an identity which exhibits the product of two sums of two squares as itself the sum of two squares. The identity is

$$(a^2+b^2)(c^2+d^2)=(ac+bd)^2+(ad-bc)^2, \tag{1}$$

and it is generally attributed to Leonardo of Pisa (also called Fibonacci), who gave it in his *Liber Abaci* of 1202.

Every number which satisfies the conditions given above can be built up as a product of factors, each of which is either 2, or a prime of the form $4k+1$, or the *square* of a prime of the form $4k+3$. If we can prove that each such factor is representable as the sum of two squares, it will follow by repeated application of the identity (1) that the number itself is representable. Now 2 is obviously representable as 1^2+1^2, and if q is a prime of the form $4k+3$ then q^2 is representable as q^2+0^2. It remains to be proved that *any prime of the form $4k+1$ is representable as x^2+y^2*, and this result will be proved in the next section. Once we have this, we have the necessary and sufficient condition for a number to be the sum of two squares, as stated above.

It must not be overlooked that in the present theory we are admitting representations by x^2+y^2 in which x and y may have a factor in common (e.g. $q^2=q^2+0^2$). If it is required that x and y shall be relatively prime, the result is slightly different. It will be found in VI.5, where the question is considered as a special case of a more general theory.

2. Primes of the form $4k+1$

We now give the classical proof, which is due essentially to Euler, that any prime p of the form $4k+1$ is representable as the sum of two squares. This proof falls into two stages. The first stage is the proof that some multiple of p is representable as z^2+1, and the second stage is the deduction from this that p itself is representable as x^2+y^2.

The first stage is equivalent to proving that the congruence

$$z^2+1\equiv0 \text{ (mod } p)$$

is soluble for any prime p of the form $4k+1$. This we already know from III.3, where the result was deduced from Euler's criterion for a number to be a quadratic residue (mod p).

The second stage of the proof starts from the fact just stated, which implies that

$$mp=z^2+1$$

for some natural number m. We can suppose that z lies between $-\frac{1}{2}p$ and $\frac{1}{2}p$, since this can be ensured by subtracting from z a suitable multiple of p. We have then

$$m = \frac{1}{p}(z^2+1) < \frac{1}{p}(\tfrac{1}{4}p^2+1) < p.$$

In order to have the argument in a form which can be applied later in more general circumstances, we shall suppose only that

$$mp=x^2+y^2 \tag{2}$$

for some integers x and y, where m is a natural number less than p. The idea of the proof is to show that if $m>1$, there is some natural number m', less than m, which has the same property. By repetition of the argument, it will eventually follow that the number 1 has the property, in other words that $p=x^2+y^2$.

The argument proceeds as follows. We determine two integers u and v which lie between $-\frac{1}{2}m$ and $\frac{1}{2}m$ (inclusive, if m is even) and which are respectively congruent to x and y to the modulus m:

$$u\equiv x, \quad v\equiv y \text{ (mod } m). \tag{3}$$

Then

$$u^2+v^2\equiv x^2+y^2\equiv0 \text{ (mod } m),$$

so that

$$mr=u^2+v^2 \tag{4}$$

for some integer r. We observe that r cannot be zero, for then u and v would be zero, so that x and y would be multiples of m, which is contrary to (2), since it would imply that the prime p was a multiple of m. As regards the magnitude of r, we have

$$r = \frac{1}{m}(u^2+v^2) \leqq \frac{1}{m}(\tfrac{1}{4}m^2+\tfrac{1}{4}m^2) < m .$$

Multiply together the two equations (2) and (4), and apply the identity (1). This gives

$$m^2 rp = (x^2+y^2)(u^2+v^2) = (xu+yv)^2+(xv-yu)^2. \tag{5}$$

The important point to be observed now is that both the numbers $xu+yv$ and $xv-yu$ are multiples of m. For, by (3),

$$xu+yv \equiv x^2+y^2 \equiv 0 \ (\text{mod } m),$$

and

$$xv-yu \equiv xy-yx \equiv 0 \ (\text{mod } m).$$

Hence the equation (5) can be divided throughout by m^2, giving

$$rp = X^2 + Y^2$$

for some integers X and Y. We have therefore proved that there is some natural number r, less than m, for which rp is representable as the sum of two squares. As explained earlier, this is enough to prove that p itself is representable.

It may be of interest to illustrate the proof by working through it in a numerical case. Take $p=277$, this being a prime of the form $4k+1$. We know that the congruence $z^2+1 \equiv 0 \ (\text{mod } 277)$ is soluble, and the solution can be found either by trial or by using a table of indices. In fact $z=60$ provides a solution, since

$$60^2+1=3601=277 \times 13.$$

Thus the starting point of the proof, analogous to (2), is

$$13 \times 277 = 60^2+1^2.$$

Following the plan of the proof, we reduce the numbers 60 and 1 to the modulus 13, obtaining the numbers -5 and 1. The equation analogous to (4) is

$$13 \times 2 = (-5)^2+1^2.$$

The next step is to multiply together the two equations, and apply the identity (1). We obtain

$$13^2 \times 2 \times 277 = (60^2 + 1^2)((-5)^2 + 1^2)$$
$$= (60 \times (-5) + 1 \times 1)^2 + (60 \times 1 - 1 \times (-5))^2$$
$$= (-299)^2 + 65^2.$$

The numbers on the right are divisible by 13, as they must be, and we obtain

$$2 \times 277 = (-23)^2 + 5^2.$$

Now we repeat the process. Reducing -23 and 5 to the modulus 2, they become 1, and the corresponding equation is

$$2 \times 1 = 1^2 + 1^2.$$

Multiplying this by the preceding equation, and applying the identity (1), we obtain

$$2^2 \times 277 = (-23 + 5)^2 + (-23 - 5)^2$$
$$= (-18)^2 + (-28)^2.$$

Hence, finally,

$$277 = 9^2 + 14^2.$$

In connection with the general theorem, there is a further remark to be made, namely that the representation of p as $x^2 + y^2$ is *unique*, apart from the obvious possibility of interchanging x and y, and changing their signs. Fermat laid stress on this fact, and called it 'the fundamental theorem on right-angled triangles', since it shows that there is exactly one right-angled triangle whose hypotenuse is $\sqrt{p}$ and whose other sides are natural numbers.

The proof of the uniqueness is not difficult. Suppose that

$$p = x^2 + y^2 = X^2 + Y^2. \tag{6}$$

We know that the congruence $z^2 + 1 \equiv 0 \pmod{p}$ has exactly two solutions, which are of the form $z \equiv \pm h \pmod{p}$. Hence

$$x \equiv \pm hy \quad \text{and} \quad X \equiv \pm hY \pmod{p}.$$

Since the signs of x, y, X, Y are immaterial, we can suppose that

$$x \equiv hy, \quad X \equiv hY \pmod{p}. \tag{7}$$

Multiply together the two equations (6), and apply the identity (1).

We obtain

$$p^2 = (x^2 + y^2)(X^2 + Y^2) = (xX + yY)^2 + (xY - yX)^2.$$

Now $xY - yX \equiv 0 \pmod p$ by (7). Hence both numbers on the right are multiples of p, and the equation can be divided by p^2 throughout. It will then reduce to an equation which expresses 1 as the sum of two integral squares, and the only possibility is $(\pm 1)^2 + 0^2$. Thus, in the previous equation, one of the two numbers $xX + yY$, $xY - yX$ must be 0. If $xY - yX = 0$ it follows, since x, y and X, Y are relatively prime, that either $x = X$ and $y = Y$ or $x = -X$ and $y = -Y$. Similarly if $xX + yY = 0$ it follows that either $x = Y$ and $y = -X$ or $x = -Y$ and $y = X$. In any case, the two representations in (6) are essentially the same.

3. Constructions for x and y

Once it was known that any prime p of the form $4k + 1$ is representable uniquely as $x^2 + y^2$, it is natural that mathematicians should have tried to find constructions for the numbers x and y in terms of p. A construction often gives greater mental satisfaction than a mere proof of existence, though the distinction between the two is not always a clear-cut one. Four constructions for x and y are known, due to Legendre (1808), Gauss (1825), Serret (1848) and Jacobsthal (1906), and we proceed to give them without entering into the details of the proofs. Part of the interest of these constructions lies in the variety of the methods which they use.

Legendre's construction is based on the continued fraction for $\sqrt{p}$. This is of the form (IV.9)

$$\sqrt{p} = q_0 + \cfrac{1}{q_1 +} \cfrac{1}{q_2 +} \cdots \cfrac{1}{q_2 +} \cfrac{1}{q_1 +} \cfrac{1}{2q_0 +} \cdots,$$

the period consisting of a symmetrical part $q_1, q_2, \cdots, q_2, q_1$ followed by $2q_0$. So far, this does not depend on p being a prime of the form $4k + 1$, and applies to any number which is not a perfect square. We recall also (IV.11) that if there is no central term in the symmetrical part of the period, then the equation $x^2 - py^2 = -1$ is soluble. The converse is also true, although it was not proved in IV.11. Legendre proved, in quite an elementary way, that if p is a prime of the form $4k + 1$, the equation $x^2 - py^2 = -1$ *is* soluble. Consequently, by the converse theorem just stated, there is no central term, and the period has the form

$$q_1, q_2, \cdots, q_m, q_m, \cdots, q_2, q_1, 2q_0.$$

Now let α be the particular complete quotient which begins at the middle of the period, that is

$$\alpha = \alpha_m = q_m + \cfrac{1}{q_{m-1}+} \cdots \cfrac{1}{q_1+} \cfrac{1}{2q_0+} \cfrac{1}{q_1+} \cdots .$$

This is a purely periodic continued fraction, whose period consists of $q_m, \cdots , q_1, 2q_0, q_1, \cdots , q_m$. Since this period is symmetrical, we have, as in IV.9, $\alpha' = -\dfrac{1}{\alpha}$, where α' denotes the conjugate of α. Now α is expressible in the form

$$\alpha = \frac{P+\sqrt{p}}{Q},$$

where P and Q are integers. The equation $\alpha\alpha' = -1$ gives

$$\frac{P+\sqrt{p}}{Q} \cdot \frac{P-\sqrt{p}}{Q} = -1,$$

or

$$p = P^2 + Q^2.$$

This is Legendre's construction.

As an illustration, take $p = 29$. The process for developing $\sqrt{29}$ in a continued fraction is

$$\sqrt{29} = 5 + \frac{1}{\alpha_1},$$

$$\alpha_1 = \tfrac{1}{4}(5+\sqrt{29}) = 2 + \frac{1}{\alpha_2},$$

$$\alpha_2 = \tfrac{1}{5}(3+\sqrt{29}) = 1 + \frac{1}{\alpha_3},$$

$$\alpha_3 = \tfrac{1}{5}(2+\sqrt{29}) = 1 + \frac{1}{\alpha_4},$$

$$\alpha_4 = \tfrac{1}{4}(3+\sqrt{29}) = 2 + \frac{1}{\alpha_5},$$

$$\alpha_5 = 5 + \sqrt{29}.$$

The continued fraction is $5, \overline{2, 1, 1, 2, 10}$. The appropriate complete quotient to take is $\alpha = \alpha_3$, giving $P = 2$ and $Q = 5$, corresponding to $29 = 2^2 + 5^2$.

The second construction is that of Gauss, and this is the most elementary of all to state, though not to prove. If $p=4k+1$, take

$$x \equiv \frac{(2k)!}{2\,(k!)^2} \pmod{p}, \quad y \equiv (2k)!\, x \pmod{p},$$

with x and y numerically less than $\frac{1}{2}p$. Then $p=x^2+y^2$. A proof was given by Cauchy, and another by Jacobsthal, but neither of these is very simple. To illustrate the construction, take again $p=29$. Then

$$x \equiv \frac{14!}{2\,(7!)^2} = 1716 \equiv 5 \pmod{29},$$
$$y \equiv 14!\, x \equiv (14!) \times 5 \equiv 2 \pmod{29}.$$

The construction is obviously not a very convenient one for purposes of numerical calculation, in spite of its elementary nature.

The third construction is that of Serret. This, like Legendre's construction, uses a continued fraction, but now the number developed is a rational number. We expand $\frac{p}{h}$ into a continued fraction, where h satisfies $h^2+1\equiv0 \pmod{p}$ and $0<h<\frac{1}{2}p$. It can be proved that the continued fraction is of the form

$$\frac{p}{h} = q_0 + \frac{1}{q_1+} \cdots \frac{1}{q_m+}\frac{1}{q_m+} \cdots \frac{1}{q_0}, \tag{8}$$

that is, the terms are symmetrical and there is no central term. With the notation of Chapter IV, let

$$x = [q_0, q_1, \ldots, q_m], \quad y = [q_0, q_1, \ldots, q_{m-1}].$$

Then

$$p = x^2 + y^2.$$

For example, if $p = 29$, we find that $h = 12$, since

$$12^2+1 = 145 = 5 \times 29.$$

The continued fraction is

$$\frac{29}{12} = 2 + \frac{1}{2+}\frac{1}{2+}\frac{1}{2}.$$

Hence

$$x = [2, 2] = 5, \quad y = [2] = 2.$$

This construction was given again in a slightly different form by H. J. S. Smith in 1855. His object was to give a simple and direct proof that any prime of the form $4k+1$ is representable as the sum of two squares. He avoided any consideration of congruences by proving directly that there is *some* number h with $0<h<\frac{1}{2}p$ for which the continued fraction for $\dfrac{p}{h}$ has the form given in (8). Defining x and y as above, he proved like Serret that $p=x^2+y^2$.

Finally, we come to Jacobsthal's construction. This is based on considerations similar to those that occurred in III.6 in connection with the distribution of quadratic residues. We consider the following sum of Legendre symbols:

$$S(a) = \sum_{n}\left(\frac{n(n^2-a)}{p}\right),$$

where a is any number not congruent to 0 (mod p), and the summation is extended over a complete set of residues, for example over $n=0, 1, 2, \ldots , p-1$. It can easily be proved that $|S(a)|$ has only two possible values, one when a is a quadratic residue, the other when a is a quadratic non-residue. Moreover, each of these values is an even integer, for the term $n=0$ contributes 0 to the sum, and two terms n and $-n$ contribute the same amount, since $(-1 \mid p)=1$. Put

$$x = \tfrac{1}{2}\,|S(R)|, \quad y = \tfrac{1}{2}\,|S(N)|,$$

where R is any quadratic residue and N any quadratic non-residue. Then

$$p = x^2 + y^2.$$

The proof is not very difficult, depending mainly on a skilful use of the relation (18) of Chapter III.

As an illustration, take $p=29$ again. For R we take 1, and for N we take 2, since this is a non-residue. The values of $n(n^2-1)$ (mod 29) consist of 0, and the numbers

$$0, 6, -5, 2, 4, 7, -12, 11, -5, 4, -14, 5, 9, 4$$

each twice. The sum of the Legendre symbols of the above numbers is 5, hence $x=5$. The values of $n(n^2-2)$ (mod 29) consist of 0 and the numbers

$$-1, 4, -8, -2, -1, 1, 10, 3, -14, -6, 4, -7, -4, -10$$

each twice. The sum of the Legendre symbols of these numbers is 2, hence $y=2$.

4. *Representation by four squares*

It was stated by Girard and by Fermat that *every natural number is representable as the sum of four squares of integers*. Another way of expressing the result (allowing for the possibility that some of the integers may be zero) is to say that every natural number is representable as the sum of at most four squares of natural numbers. Some historians have argued that the fact was known already to Diophantus of Alexandria, because he made no mention of any condition to be satisfied by a number for it to be representable as a sum of four squares, whereas he was aware that only certain kinds of numbers could be represented by two or three squares.

Euler made many attempts to prove the result, but did not succeed. His failure may have been due to the fact that he tried to represent the given number as the sum of two numbers, each of which satisfies the conditions for representation by two squares. Such an approach to the question does not easily lead to a proof. The first proof was given in 1770 by Lagrange, who acknowledged his great indebtedness to the work of Euler.

Lagrange's proof is very similar to that given in §§1 and 2 for the result concerning two squares, apart from one slight complication. Again there is an identity which expresses the product of two sums of four squares as itself the sum of four squares. This identity (due to Euler) is as follows:

$$\begin{cases} (a^2+b^2+c^2+d^2)(A^2+B^2+C^2+D^2) \\ \quad = (aA+bB+cC+dD)^2+(aB-bA-cD+dC)^2 \\ \quad +(aC+bD-cA-dB)^2+(aD-bC+cB-dA)^2. \end{cases} \quad (9)$$

In view of this identity, it suffices to prove that every *prime* is representable as the sum of four squares, for then the representability of composite numbers will follow by repeated application of the identity. Since we already know that the prime 2 and all primes of the form $4k+1$ are representable by two squares, it remains only to prove that *any prime of the form $4k+3$ is representable as the sum of four squares*.

The proof falls into two stages, like that in §2. The first stage is the proof that some multiple mp of p, where $0<m<p$, is representable as the sum of four squares. The second stage is the deduction from this that p itself is representable.

For the first stage it is enough to prove that the congruence

$$x^2 + y^2 + 1 \equiv 0 \ (\mathrm{mod}\ p) \tag{10}$$

is soluble. For then we can choose a solution with x and y each numerically less than $\frac{1}{2}p$, and we have

$$mp = x^2 + y^2 + 1^2 + 0^2,$$

with

$$m < \frac{1}{p}\left(\tfrac{1}{4}p^2 + \tfrac{1}{4}p^2 + 1\right) < p.$$

Euler gave a simple argument which establishes the solubility of the congruence (10) without any calculation. We write the congruence as

$$x^2 + 1 \equiv -y^2 \ (\mathrm{mod}\ p).$$

Any quadratic non-residue (mod p) is representable as congruent to some number of the form $-y^2$, since -1 is a quadratic non-residue for any prime of the form $4k+3$ (III.3). Thus, to satisfy the above congruence, it suffices to find a quadratic residue R and a quadratic non-residue N such that $R+1=N$. If we take N to be the *first* quadratic non-residue in the series 1, 2, 3, ... , this condition is obviously satisfied, and the solubility of the congruence follows.

We may observe in passing that the solubility of the congruence (10) is a special case of Chevalley's theorem (II.8). We saw there that the congruence

$$x^2 + y^2 + z^2 \equiv 0 \ (\mathrm{mod}\ p)$$

is soluble, with not all of x, y, z congruent to 0. If we suppose $z \not\equiv 0$, and determine X and Y so that $x \equiv Xz$, $y \equiv Yz$, we then have $X^2 + Y^2 + 1 \equiv 0$.

We now come to the second stage of the proof, which starts from the fact that mp is representable as

$$mp = a^2 + b^2 + c^2 + d^2, \tag{11}$$

for some number m with $0 < m < p$. We shall prove, in almost the same way as in §2, that if $m > 1$ there is some number r with $0 < r < m$ which has the same property as m. It follows, by repetition of the argument, that the number 1 has the property, and therefore that p itself is representable as the sum of four squares.

We begin by reducing a, b, c, d with respect to the modulus m, that is, we determine numbers A, B, C, D which are respectively congruent to a, b, c, d to the modulus m, and which satisfy $-\frac{1}{2}m < A \leqq \frac{1}{2}m$, and so on for B, C, D. We now have

$$mr = A^2 + B^2 + C^2 + D^2 \qquad (12)$$

for some integer r. This number r cannot be zero, for then A, B, C, D would all be zero, and a, b, c, d would all be multiples of m. From (11) we would have mp divisible by m^2, or p divisible by m, which is impossible since p is a prime and m is greater than 1 but less than p.

As regards the magnitude of r, we have

$$r = \frac{1}{m}(A^2 + B^2 + C^2 + D^2) \leqq \frac{1}{m}(\tfrac{1}{4}m^2 + \tfrac{1}{4}m^2 + \tfrac{1}{4}m^2 + \tfrac{1}{4}m^2) = m.$$

This is not good enough as it stands; we need to know that r is strictly *less* than m. The possibility that $r = m$ will only arise if A, B, C, D are all equal to $\frac{1}{2}m$. In this case m is even, and A, B, C, D are all congruent to $\frac{1}{2}m$ to the modulus m. But then $a^2 \equiv \frac{1}{4}m^2$ (mod m^2), and similarly for b, c, d. Now (11) gives $mp \equiv 0$ (mod m^2) and, as we have already seen, this is impossible. It follows that the number r in (12) satisfies $0 < r < m$.

We continue the proof by multiplying together the equations (11) and (12), and applying the identity (9). This gives

$$m^2 rp = x^2 + y^2 + z^2 + w^2, \qquad (13)$$

where x, y, z, w are the four expressions on the right-hand side of (9). All these expressions represent numbers which are divisible by m. For

$$x = aA + bB + cC + dD \equiv a^2 + b^2 + c^2 + d^2 \equiv 0 \text{ (mod } m),$$

and

$$y = aB - bA - cD + dC \equiv ab - ba - cd + dc \equiv 0 \text{ (mod } m),$$

with similar results for z and w. We can cancel m^2 from both sides of the equation (13), and obtain a representation for rp as the sum of four squares. This proves the desired result.

The above proof of Lagrange's four-square theorem is a little simpler than the proof he originally gave, and is essentially that given later by Euler. Although the details of the proof can be varied somewhat, I do not know of any other simple and elementary proof which is fundamentally different from this one.

5. Representation by three squares

This is a much more difficult question. One reason for the difficulty lies in the fact that there is no such identity as (1) or (9). Indeed, it is very easy to see that the product of two numbers, each a sum of three squares, need not itself be a sum of three squares. For example, $3 = 1^2 + 1^2 + 1^2$ and $5 = 2^2 + 1^2 + 0^2$, but 15 is not representable as the sum of three squares.

As in §1, we can rule out some numbers as incapable of being represented as a sum of three squares. Any square is congruent to 0 or 1 or 4 to the modulus 8. Hence the sum of three squares cannot be congruent to 7 (mod 8), since it is impossible to build up 7 from three terms, each of which is 0 or 1 or 4. Hence a number of the form $8k + 7$ is not representable.

Further, a multiple of 4, say $4m$, can only be representable if m itself is representable. For any square is congruent to 0 or 1 (mod 4), and the sum of three squares can only be divisible by 4 if all the numbers are even. Hence numbers of the form $4(8k + 7)$ are not representable, and numbers of the form $16(8k + 7)$ are not representable and so on. In general, we can say that a number of the form $4^l(8k + 7)$ is not representable as a sum of three squares.

It is a fact that every number which is *not* of this form *is* representable. The first proof was attempted by Legendre, but in the course of it he assumed that any arithmetical progression a, $a + b$, $a + 2b$, ... (in which a and b are relatively prime) must contain infinitely many primes. This was first proved by Dirichlet in 1837, forty years after Legendre's work. Gauss, in his *Disquisitiones Arithmeticae*, gave a complete proof, but it was one which depended on the more difficult results in his extensive theory of quadratic forms. Other proofs have since been given, but none of them can be described as both elementary and simple.

NOTES

§1. The reader who is familiar with complex numbers will recognize the identity (1) as equivalent to $|\alpha\beta|^2 = |\alpha|^2|\beta|^2$, where $\alpha = a + ib$ and $\beta = c + id$. The numbers of the form $a + ib$, where a and b are integers, are the so-called *Gaussian integers*, and to represent n as the sum of two squares is the same thing as to find Gaussian integers $a + ib$ whose norm $a^2 + b^2$ is n. The theory takes on a more elegant appearance when it is expressed in terms of Gaussian integers.

§3. For references, see Dickson's *History*, vol. II, ch. 6 and vol. III, ch. 2. The various constructions do not generally give positive values for *x* and *y*, though they happen to do so when *p* is 29.

§4. The identity (9) bears the same relation to quaternions as the identity (1) bears to complex numbers (see note to §1 above). Hurwitz gave a treatment of representation by four squares by means of quaternions; for an account see Hardy and Wright, ch. 20.

§5. A proof of the three squares theorem, based on Dirichlet's theorem on primes in arithmetical progressions, is given in Landau's *Vorlesungen über Zahlentheorie*, vol. I, pp. 114–21.

Rational squares. It follows from the condition (§1) for a number to be a sum of two squares that *if an integer is expressible as the sum of two rational squares then it is expressible as the sum of two integral squares.* Similarly for three squares, in view of the condition given in §5.

Number of representations. Lack of space precludes us from giving an account of the formulae that are known for the number of representations of a number n as the sum of two squares, or four squares. In these formulae, the representations are supposed to be by integers, which may be positive, negative or zero, and two representations are counted as distinct unless they are identical. For two squares the rule (due to Legendre) is as follows. Count the number of divisors of n of the form $4x + 1$ and the number of those of the form $4x + 3$. If these numbers are D_1 and D_3 respectively, then the number of representations is $4(D_1 - D_3)$. For four squares, the rule was found by Jacobi, who deduced it from an identity connecting two infinite series. If n is odd, the number of representations of n as the sum of four squares is $8\sigma(n)$. If n is even, put $n = 2^r n'$ where n' is odd; then the number of representations of n is $24\sigma(n')$. Here $\sigma(n)$ denotes the sum of the divisors of n, as in I.5. For proofs of these results see, for example, Hardy and Wright, chs. 16, 20. The number of representations by three squares is a much more recondite function, but can be expressed in terms of certain class-numbers of quadratic forms (VI.9).

VI

QUADRATIC FORMS

1. *Introduction*

In Chapter V we found the necessary and sufficient condition for a number to be representable as the sum of two squares, the condition being one that related to the prime factors of the number. Euler and other mathematicians of the eighteenth century were also successful in finding the necessary and sufficient conditions for a number to be representable as x^2+2y^2 or x^2+3y^2, and again these related to the prime factors of the number. It was natural that they should then try to find similar results for general quadratic forms. A quadratic form, in this connection, means an expression

$$ax^2 + bxy + cy^2$$

which is homogeneous and of the second degree in the variables, and has integral coefficients a, b, c. We shall limit ourselves to forms in two variables, or *binary* forms, though there is also a theory of quadratic forms in three variables (ternary forms), or in any number of variables.

The theory of quadratic forms was first developed by Lagrange in 1773, and many of the fundamental ideas are due to him. His theory was simplified and extended by Legendre, and further progress was made by Gauss, who introduced many new concepts and used them to prove deep and difficult results which had eluded Lagrange and Legendre.

The classical problem of the subject is the *problem of representation*: given a particular quadratic form, what are the numbers represented by it? A simple answer can be given for some special forms, such as x^2+y^2 or x^2+2y^2 or x^2+3y^2; but there is no such simple answer in the general case. What the theory does lead to is a simple answer to a rather different problem: that of representation not by one form but by one or other of a certain set of forms.

The general ideas of the theory, which all arise out of the notion of equivalence (§2), are of importance in other more difficult and more advanced theories. The study of quadratic forms provides a natural introduction to them, and allows one to become familiar with them in a context where they are readily appreciated.

2. *Equivalent forms*

A fundamental notion in connection with quadratic forms (and other forms, too) is that of equivalence. We recognize at once that two forms such as $2x^2+3y^2$ and $3x^2+2y^2$ are really the same, one being obtained from the other by merely interchanging the variables. It is not quite so obvious that the form $2x^2+4xy+5y^2$ is essentially the same as either of the two forms just mentioned. However, this form can be written as

$$2(x+y)^2 + 3y^2,$$

and when the variables x and y take all integral values, so do the variables $x+y$ and y, and conversely. It is clear that any property of a general nature possessed by the form $2x^2+3y^2$ will also be possessed by the form $2(x+y)^2+3y^2$, and conversely. Certainly this is true of properties relating to the representation of numbers: if we know the representations of a number by one of the forms then we can immediately deduce what are the representations by the other. The two forms are connected by a very simple *substitution*: if we put $x=X+Y$ and $y=Y$, then

$$2x^2 + 3y^2 = 2X^2 + 4XY + 5Y^2.$$

This substitution has the property that as x and y take all integral values, so also do X and Y, and conversely.

We ask ourselves the general question: what substitutions of the form

$$x = pX + qY, \quad y = rX + sY \tag{1}$$

have this property, that is, establish a one-to-one correspondence between all integer pairs x, y and all integer pairs X, Y? We do not impose *a priori* any restriction on the nature of the coefficients

p, q, r, s, though in fact it is obvious that they must all be integers, for the values $x=p$, $y=r$ correspond to the values $X=1$, $Y=0$, and the values $x=q$, $y=s$ correspond to the values $X=0$, $Y=1$. If all four coefficients are integers, then whatever integral values we give to X and Y, the resulting values of x and y will be integers.

We want the converse to be true also. The obvious way to investigate this is to express X and Y in terms of x and y. If we multiply the first equation by s and the second by q and subtract, we obtain

$$sx - qy = (ps - qr) X,$$

and in a similar way we get

$$-rx + py = (ps - qr) Y.$$

The number $ps - qr$ cannot be zero, for then $sx - qy$ and $-rx + py$ would always be zero, and the variables x and y would not be independent. Putting $\Delta = ps - qr$, and dividing by Δ, the equations expressing X and Y in terms of x and y are

$$X = \frac{s}{\Delta} x - \frac{q}{\Delta} y, \quad Y = -\frac{r}{\Delta} x + \frac{p}{\Delta} y. \tag{2}$$

The four coefficients here must also be integers. This is certainly true if $\Delta = \pm 1$. It will not be true otherwise; for if the four coefficients are integers, then so also is

$$\frac{p}{\Delta} \frac{s}{\Delta} - \frac{q}{\Delta} \frac{r}{\Delta},$$

and the value of this is $\frac{1}{\Delta}$, which is only an integer if $\Delta = \pm 1$.

Hence *the coefficients p, q, r, s of the substitution must all be integers, and $ps - qr$ must be ± 1.* Then, and only then, will the substitution have the desired property of making all integer pairs x, y correspond to all integer pairs X, Y, and vice versa.

The expression $ps - qr$ is called the *determinant* of the substitution. In order to avoid complications in the later theory, it is customary to restrict oneself to the use of substitutions of determinant 1, and to make no use of those of determinant -1. A substitution of the form (1) with integral coefficients and determinant 1 will be called a *unimodular* substitution.

Two forms which are related by a unimodular substitution are said to be *equivalent*. For example, as we saw above, the form

$2x^2+3y^2$ can be transformed into the form $2X^2+4XY+5Y^2$ by the substitution

$$x = X + Y, \quad y = Y,$$

which is a unimodular substitution, and so the two forms are equivalent. To avoid specifying particular letters for the variables, and changing them at each substitution, it is convenient to denote the quadratic form $ax^2+bxy+cy^2$ by (a, b, c), and to express the equivalence of two forms by the symbolism

$$(2, 0, 3) \sim (2, 4, 5).$$

The original example was

$$(2, 0, 3) \sim (3, 0, 2),$$

but here one comment must be made. The substitution which interchanges the variables, that is, the substitution $x=Y, y=X$, is not a unimodular substitution according to our present definition, because its determinant is -1. Instead, however, we can use the substitution $x=Y, y=-X$, which is unimodular, and transforms $(2, 0, 3)$ into $(3, 0, 2)$. Applied to a general form, this substitution shows that

$$(a, b, c) \sim (c, -b, a). \tag{3}$$

In using the term 'equivalence', we have been tacitly assuming that this relationship between two forms has certain simple properties; if this were not so, the use of the word would be misleading. The properties are: (i) any form is equivalent to itself, (ii) if one form is equivalent to another, then the second form is equivalent to the first, (iii) two forms which are equivalent to the same form are equivalent to one another. In fact, all these properties follow at once from the definition. First, any form is equivalent to itself by the *identical substitution* $x=X, y=Y$. Secondly, if one form is transformed into another by the substitution (1), then the second form will be transformed back into the first by the inverse substitution (2), where now $\Delta=1$. Finally, the third result follows from the fact that two unimodular substitutions applied in succession can be replaced by a single unimodular substitution. If the substitution

$$x = pX + qY, \quad y = rX + sY$$

is followed by the substitution

$$X = P\xi + Q\eta, \quad Y = R\xi + S\eta,$$

the final effect is the same as that of the substitution

$$x = p\,(P\xi + Q\eta) + q\,(R\xi + S\eta),$$
$$y = r\,(P\xi + Q\eta) + s\,(R\xi + S\eta).$$

This resultant substitution has integral coefficients, and its determinant is

$$(pP + qR)(rQ + sS) - (pQ + qS)(rP + sR) = (ps - qr)(PS - QR),$$

and so is 1.

It is obvious (as we have already remarked in a particular case) that the problem of representation is the same for two equivalent forms. A similar remark applies to a modified form of the problem: that of *proper* representation. A number n is said to be properly representable by a form (a, b, c) if $n = ax^2 + bxy + cy^2$, where x and y are *relatively prime* integers. A unimodular substitution transforms relatively prime pairs x, y into relatively prime pairs X, Y, and conversely; for if X and Y had a common factor in (1), then x and y would have the same common factor. It follows that if two forms are equivalent, the proper representations of a number by the two forms correspond to one another by the unimodular substitution.

3. The discriminant

The discriminant of a quadratic form (a, b, c) is defined to be the number $b^2 - 4ac$. Thus the discriminant of the form $(2, 0, 3)$ is -24, and the discriminant of the form $(2, 4, 5)$ is $4^2 - 4 \times 2 \times 5 = -24$ also.

It is an important fact that *equivalent forms have the same discriminant*. The shortest proof is by direct verification. If we apply the substitution (1) to the form $ax^2 + bxy + cy^2$ we get the form $AX^2 + BXY + CY^2$, where

$$\begin{cases} A = ap^2 + bpr + cr^2, \\ B = 2apq + b(ps + qr) + 2crs, \\ C = aq^2 + bqs + cs^2. \end{cases} \tag{4}$$

It can be verified that

$$B^2 - 4AC = (b^2 - 4ac)(ps - qr)^2. \tag{5}$$

Since $ps - qr = 1$, the two forms (a, b, c) and (A, B, C) have the same discriminant. Naturally, the identity (5) does not depend on the nature of the coefficients p, q, r, s in the substitution. It is a purely algebraical relation, and we have here a particular instance of a very general situation. A function of the coefficients of an algebraic form, such as $b^2 - 4ac$ in the present case, which is unaltered when a linear substitution of determinant 1 is applied to the form, is said to be an *algebraic invariant* of the form. The discriminant of a binary quadratic form is a simple example of such an invariant.

Although equivalent forms have the same discriminant, it is by no means true that forms of the same discriminant are necessarily equivalent. For example, the forms (1, 0, 6) and (2, 0, 3) both have the discriminant -24, but they are not equivalent. To see this, we need only observe that the form $x^2 + 6y^2$ represents the number 1, namely when $x = 1$ and $y = 0$, whereas the form $2x^2 + 3y^2$ can obviously never take the value 1.

The discriminant d of a quadratic form is an integer, which may be positive, negative or zero. Not every integer can figure as the discriminant of a form. For $b^2 - 4ac \equiv b^2 \pmod 4$, and any square is congruent to 0 or 1 (mod 4). Hence d must be congruent to 0 or 1 (mod 4), and the possible discriminants are

$$\ldots, -11, -8, -7, -4, -3, 0, 1, 4, 5, 8, 9, \ldots.$$

Moreover, each such number *is* the discriminant of at least one form. For if d is any given number which is congruent to 0 or 1 (mod 4), we can satisfy the equation $b^2 - 4ac = d$ by taking a to be 1, and taking b to be 0 or 1 according as $d \equiv 0$ or 1 (mod 4). Then c is $-\frac{1}{4}d$ or $-\frac{1}{4}(d-1)$, as the case may be. This gives a particular form of discriminant d, namely

$$(1, 0, -\tfrac{1}{4}d) \quad \text{or} \quad (1, 1, -\tfrac{1}{4}(d-1))$$

according as $d \equiv 0$ or 1 (mod 4). This is called the *principal form* of discriminant d. Thus the principal form of discriminant -4 is (1, 0, 1), or $x^2 + y^2$, and the principal form of discriminant 5 is (1, 1, -1), or $x^2 + xy - y^2$.

There is an important distinction to be made between forms of positive discriminant and forms of negative discriminant. (We shall not consider forms of zero discriminant, since such a form is simply the square of a certain linear form.) Let us first consider forms of *negative* discriminant. We multiply the form by $4a$ and carry out the process of 'completing the square', as follows:

$$4a\,(ax^2 + bxy + cy^2) = 4a^2x^2 + 4abxy + 4acy^2$$
$$= (2ax + by)^2 + (4ac - b^2)y^2.$$

Here $4ac-b^2$ is positive. Hence the last expression is always positive, whatever values x and y may have, except that it is zero when x and y are both zero. It follows that all the numbers represented by the form have the same sign: they are all positive if a is positive, or all negative if a is negative. Such a form is said to be *definite*, and to be *positive definite* or *negative definite* as the case may be. We can always change a negative definite form into a positive definite form by merely changing the signs of all the coefficients, and therefore in treating definite forms it is enough to consider positive definite forms. Examples of positive definite forms are $(1, 3, 7)$, of discriminant -19, or $(5, -7, 5)$, of discriminant -51.

Consider next forms of *positive* discriminant. The expression obtained above is still valid, but since $4ac-b^2 = -d$, and d is now positive, we can factorize it. We obtain

$$4a(ax^2 + bxy + cy^2) = (2ax + by + \sqrt{d}y)(2ax + by - \sqrt{d}y)$$
$$= 4a^2(x - \theta y)(x - \phi y),$$

where θ and ϕ are given by

$$\frac{-b \pm \sqrt{d}}{2a}.$$

Here we assume, for the moment, that a is not zero. The numbers θ and ϕ are real, but not generally rational. The sign of the product $(x-\theta y)(x-\phi y)$ depends on whether the fraction $\frac{x}{y}$ falls between the two numbers θ and ϕ, or outside them. Since there are fractions of both kinds, the form assumes both positive and negative values. It is said to be *indefinite*. The case when a is zero is still simpler; here the form factorizes as $y(bx+cy)$, and obviously takes both positive and negative values. Examples of indefinite forms are $(3, 1, -1)$, of discriminant 13, or $(1, 4, 1)$, of discriminant 12. Note that, as in the last example, the fact that the coefficients are all positive does not prevent the form from being indefinite.

We have now seen that forms of negative discriminant are definite, and forms of positive discriminant are indefinite. The first stage of the theory now to be expounded, in which the problem of representation is reduced to the problem of equivalence, applies equally to definite and indefinite forms. The later theory takes quite different shapes for the two types of form, and owing to limitations of space we shall then have to restrict ourselves almost entirely to definite forms.

4. The representation of a number by a form

In discussing what numbers are represented by a given form (a, b, c) it is enough to consider proper representation. When we know what numbers are properly representable, we can deduce what numbers are improperly representable by multiplying throughout by any square.

Suppose a number n is properly representable by a form (a, b, c). For a reason which will appear in a moment, we denote by p and r the integers for which the representation takes place, so that

$$n = ap^2 + bpr + cr^2, \qquad (6)$$

and p, r are relatively prime. If the form is definite, say positive definite, we suppose n to be positive, but if the form is indefinite n may be positive or negative. But we shall suppose that n is not zero, as that possibility is best dealt with separately (and is of little interest).

Since p and r are relatively prime, we can find integers q and s for which $ps - qr = 1$. Now consider the effect of applying the unimodular substitution (1), with these particular coefficients p, q, r, s, to the form (a, b, c). On comparing (6) with the first formula in (4), we see that the first coefficient of the transformed form is n. So we get a form, say (n, h, l), which is equivalent to the form (a, b, c), and has n as its first coefficient. Conversely, if there exists such a form, then n is properly represented by it (namely when $X = 1$ and $Y = 0$), and is therefore properly representable by the form (a, b, c). The conclusion is that *the numbers that are properly representable by a form (a, b, c) are precisely those numbers which figure as first coefficients of forms equivalent to (a, b, c).*

At first sight it may seem that this method of attacking the problem is not likely to get one very far; nevertheless it is the basis on which the whole theory rests. The problem of representation is now reduced to the problem of equivalence, in the sense that we now wish to be able to decide whether any form with the given first coefficient n is equivalent to the given form (a, b, c).

There is a simple but important deduction to be made from the general principle enunciated above. A form (n, h, l) cannot be equivalent to the given form (a, b, c) unless the two have the same discriminant, that is,

$$h^2 - 4nl = d, \tag{7}$$

where $d = b^2 - 4ac$ is the discriminant of the given form. In other words, there must exist a number h for which $h^2 - d$ is a multiple of $4n$. That is, the congruence

$$h^2 \equiv d \,(\text{mod }4n'), \quad \text{where } n' = |\,n\,|, \tag{8}$$

must be soluble. (We have to take $4n'$ as the modulus of the congruence rather than $4n$, since n may be negative.) The converse is only true to a limited extent. If the congruence (8) is soluble, there is some form (n, h, l) which has the discriminant d, but this form need not be equivalent to the given form (a, b, c). The conclusion therefore is that *if n is properly representable by any form of discriminant d, the congruence* (8) *is soluble. Conversely, if the congruence is soluble then n is properly representable by some form of discriminant d.*

In several simple cases it happens that all forms of discriminant d are mutually equivalent. In such a case, the solubility of the congruence is the necessary and sufficient condition for n to be properly representable by the given form. In the next section we apply the above principle in three such cases.

But before going on to this, there is one further remark we should make. The general principle stated above requires us to solve the congruence (8), and then to decide whether or not the resulting form (n, h, l), where l is found from $h^2 - 4nl = d$, is equivalent to the given form (a, b, c). As it stands, this might involve an infinity of trials, one for each number h which satisfies (8). In fact, however, it is enough to consider values of h which satisfy

$$0 \leqq h < 2n'. \tag{9}$$

For if h is any solution of the congruence, and (n, h, l) the corresponding form, we can apply to this form the special substitution

$$x = X + uY, \quad y = Y,$$

where u is any integer. This gives the form

$$n(X + uY)^2 + h(X + uY)Y + lY^2.$$

The first coefficient is still n, and the middle coefficient, instead of being h, is now $h+2un$. Consequently two forms with first coefficient n and with middle coefficients which differ by a multiple of $2n$ are necessarily equivalent. So it is enough to consider those forms for which h satisfies the inequality (9) as well as the congruence (8).

5. Three examples

Consider first the form x^2+y^2, of discriminant -4. It will be proved in §7 that all forms of discriminant -4 are mutually equivalent. Assuming this, the general principle tells us that a positive integer n is properly representable by x^2+y^2 if and only if the congruence

$$h^2 \equiv -4 \pmod{4n}$$

is soluble. Since h must be even to satisfy such a congruence, we can divide by 4, and consider instead the congruence

$$h^2 \equiv -1 \pmod{n}. \tag{10}$$

The question of the solubility of such a congruence is obviously bound up with the theory of quadratic residues. In the first place, by the general principle governing congruences to a composite modulus (II.6), it suffices to decide whether the congruence

$$h^2 \equiv -1 \pmod{p^r} \tag{11}$$

is soluble, for each prime power p^r occurring in the factorization of n.

The congruence (11) cannot be soluble if p is of the form $4k+3$, since -1 is a quadratic non-residue to such a modulus (III.3). If p is a prime of the form $4k+1$, the congruence is known to be soluble when r is 1, since -1 is a quadratic residue to such a modulus. It is easy to prove by induction that it is then soluble for any exponent r. For example, if r is 2, we take a number h_1 which satisfies $h_1^2 \equiv -1 \pmod{p}$, and then try to satisfy $h^2 \equiv -1 \pmod{p^2}$ by taking h to be h_1+tp, where t is an unknown. With this value of h,

$$h^2 + 1 = h_1^2 + 1 + 2th_1p + t^2p^2.$$

This will be divisible by p^2 if

$$\frac{1}{p}(h_1^2 + 1) + 2th_1 \equiv 0 \pmod{p},$$

where the first term is an integer by hypothesis. This is a linear congruence for t, and is soluble because $2h_1$ is not congruent to 0 (mod p). The same argument continues to apply for higher exponents; to solve the congruence when r is 3 we take a number h_2 which satisfies $h_2{}^2 \equiv -1$ (mod p^2) and put $h = h_2 + tp^2$, getting again a linear congruence for t to the modulus p.

This settles the question of the solubility of the congruence (11) for primes of the form $4k+1$ and $4k+3$. There remains the prime 2. Here the congruence when $r=1$ is obviously soluble (a solution being $h=1$). But it is not soluble when $r \geqq 2$, for any square is congruent to 0 or 1 (mod 4), and so cannot be congruent to -1 (mod 2^r) if $r \geqq 2$.

The conclusion therefore is that the congruence (10) is soluble if and only if *n has no prime factor of the form* $4k+3$ *and is also not divisible by* 4. This, then, is the necessary and sufficient condition for n to be properly representable as $x^2 + y^2$. If we allow for multiplication by any square, we obtain again the condition already found in Chapter V for a number to be representable as the sum of two squares, whether properly or improperly.

As a second example, take the form $x^2 + xy + 2y^2$, of discriminant -7, again a positive definite form. It will be proved in §7 that all forms of discriminant -7 are mutually equivalent. Assuming this, we have to decide for what numbers n the congruence

$$h^2 \equiv -7 \ (\text{mod } 4n) \tag{12}$$

is soluble. For simplicity we shall suppose that n is odd, so that 4 and n are relatively prime. The congruence $h^2 \equiv -7$ (mod 4) is certainly soluble, e.g. by $h=1$. The congruence $h^2 \equiv -7$ (mod p) is soluble for a prime p if -7 is a quadratic residue (mod p). The law of quadratic reciprocity (III.5) tells us which primes have this property. Provided p is not 7, we have

$$\left(\frac{-7}{p}\right) = \left(\frac{-1}{p}\right)\left(\frac{7}{p}\right) = \left(\frac{-1}{p}\right)(-1)^{\frac{1}{2}(p-1)}\left(\frac{p}{7}\right) = \left(\frac{p}{7}\right),$$

and this is $+1$ if p is of the form $7k+1$ or $7k+2$ or $7k+4$ and -1 if p is of the form $7k+3$ or $7k+5$ or $7k+6$. Exactly as before, one can prove that if the congruence is soluble for a prime modulus it is soluble for every power of that prime. There remains the case $p=7$. Here the congruence $h^2 \equiv -7$ (mod 7) is obviously soluble ($h=0$), but the congruence $h^2 \equiv -7$ (mod 7^2) is not soluble. The conclusion is that the congruence (12) is soluble if and only if *n has no prime factor of the form* $7k+3$ *or* $7k+5$ *or* $7k+6$ *and also is not divisible by* 49. This, then, is the necessary and sufficient condition for an odd number n to be properly representable as $x^2+xy+2y^2$.

As a final illustration, we take the indefinite form x^2-2y^2, of discriminant 8. Again it is true that all forms of discriminant 8 are mutually equivalent, though we shall not prove this. The congruence to be considered is

$$h^2 \equiv 8 \pmod{4n'}, \quad \text{where} \quad n' = |n|,$$

which can equally well be replaced by

$$h^2 \equiv 2 \pmod{n'}.$$

We find that the congruence $h^2 \equiv 2 \pmod{p^r}$ is soluble if p is a prime of the form $8k+1$ or $8k-1$, but not if p is a prime of the form $8k+3$ or $8k-3$. If p is 2, the congruence is soluble if $r=1$ but not if $r \geq 2$. The conclusion is that *for a number n* (*positive or negative*) *to be properly representable by* x^2-2y^2, *the criterion is that* $|n|$ *must not have any prime factor of the form* $8k+3$ *or* $8k-3$ *and must not be divisible by* 4.

Of course, it will not always happen that the condition for representation by an indefinite form is one that involves only $|n|$, and so is the same for n and $-n$. The reason why it happens here is that the form x^2-2y^2 is equivalent to the form $-x^2+2y^2$, and this is implicit in the fact that all forms of discriminant 8 are mutually equivalent.

6. *The reduction of positive definite forms*

All the infinitely many forms of a given discriminant d can be distributed into classes by placing any two equivalent forms in the same class. If this is done, two forms of discriminant d will be equivalent if and only if they belong to the same class. As we shall see later, there are only a finite number of these classes.

Given any form, it is obviously desirable to find, among the forms equivalent to it, one which is as simple as possible, using the word 'simple' as a vague term to be made precise later. This aim is achieved by the theory of reduction. As the theory takes different shapes according as it relates to definite or indefinite forms, we shall now restrict ourselves to definite forms. The theory of the reduction of indefinite forms is more difficult, and considerations of space will preclude us from giving any account of it.

The theory of the reduction of positive definite forms is due to

Lagrange. We observe first that a and c are positive for such a form, whereas b may be positive or negative. We concentrate our attention on a and $|b|$, and consider two operations of equivalence by which it may be possible to diminish one of these without altering the other. These operations are:

(i) if $c<a$, replace (a, b, c) by the equivalent form $(c, -b, a)$;

(ii) if $|b|>a$, replace (a, b, c) by the equivalent form (a, b_1, c_1), where $b_1=b+2ua$, and the integer u is so chosen that $|b_1|\leqq a$, and c_1 is then found from the fact that $b_1^2-4ac_1=d$.

The equivalence in (i) is by the substitution $x=Y, y=-X$, and the equivalence in (ii) is by the substitution $x=X+uY, y=Y$, used at the end of §4.

In operation (i), we diminish a without changing the value of $|b|$, and in operation (ii) we diminish $|b|$ without changing the value of a. Given any form, we can apply these operations alternately until we reach a form which does not satisfy either of the hypotheses for the two operations, and obviously such a form must be reached in a finite number of steps. For such a form, we have

$$c \geqq a \quad \text{and} \quad |b| \leqq a. \tag{13}$$

We have therefore proved that any positive definite form is equivalent to one whose coefficients satisfy the conditions (13).

As an illustration, we apply the process to the form $(10, 34, 29)$ of discriminant -4. Since $b>a$, we use operation (ii) to reduce b to lie in the interval from -10 to 10 by subtracting the appropriate multiple of 20, in this case 40. This gives the form $(10, -6, ?)$, and the missing coefficient is found from the discriminant. If c_1 is the new third coefficient, we have $(-6)^2-40c_1=-4$, whence $c_1=1$. The new form is $(10, -6, 1)$, and to this we apply operation (i), getting the form $(1, 6, 10)$. Now apply (ii), which in this case allows us to reduce the middle coefficient to zero. This gives the form $(1, 0, ?)$, and the missing third coefficient is found from the discriminant to be 1. Finally, we have proved that the given form is equivalent to $(1, 0, 1)$.

At the start of this process, it may happen that the given form satisfies the conditions for applying both the operations (i) and (ii). For example, if the given form is $(15, 17, 10)$ we can begin either by applying (i), obtaining $(10, -17, 15)$, or by applying (ii), obtaining $(15, -13, 8)$.

Returning to the inequalities (13), we observe that there are two cases in which, even though these inequalities hold, we may be able to apply one of the operations to some useful effect. First, if $b=-a$ we can apply operation (ii) and change b into $+a$. Secondly, if $c=a$ we can apply operation (i) and change the sign of b, thus ensuring that b is positive or zero. When we take these two possibilities into account, it follows that *any positive definite form is equivalent to one whose coefficients satisfy*

$$\begin{cases} either \quad c > a \quad and \quad -a < b \leqq a, \\ or \quad\quad c = a \quad and \quad\;\; 0 \leqq b \leqq a. \end{cases} \tag{14}$$

Such a form is called a *reduced* form.

It is a remarkable and important theorem that there is one *and only one* reduced form equivalent to a given form. The proof, though not very difficult, depends on arguments rather more elaborate than those used above. The essential idea of the proof is that of finding invariant interpretations for the coefficients of a reduced form, that is, interpretations which show that the reduced form equivalent to a given form is unique. For example, it can be proved that the first coefficient a of a reduced form is the least number which is properly represented by the form. But as the proof would take some space to set out in detail, we shall not give it here.

In view of this theorem, the question whether two given forms are equivalent or not can be answered, in any particular case, by reducing each of the forms. If the two reduced forms are the same then the two given forms are equivalent, otherwise not.

7. The reduced forms

It is easy to deduce from the inequalities (14) that there are only a finite number of reduced forms of a given negative discriminant d. Put $d=-D$, so that D is positive and

$$4ac - b^2 = D. \tag{15}$$

Since $b^2 \leqq a^2 \leqq ac$ by (14), we have $3ac \leqq D$. There are only a finite number of positive integers a and c satisfying this condition, and for each choice of a and c there are at most two possibilities for b, from (15). Hence the result. The number of reduced forms is of course the same as the number of classes of equivalent forms, since there is just one reduced form in each class. This number is called the *class-number* of the discriminant d.

To enumerate the reduced forms for a given discriminant, perhaps the quickest way is to start from the fact that

$$b^2 \leq ac \leq \tfrac{1}{3} D$$

and that $4ac = D + b^2$. Also b must be even if $D \equiv 0$ (mod 4) and odd if $D \equiv 3$ (mod 4), corresponding to $d \equiv 1$ (mod 4). One gives b all values of the appropriate parity (positive and negative) up to $\sqrt{\tfrac{1}{3}D}$, and factorizes $\tfrac{1}{4}(D + b^2)$ into ac in every possible way, and then one rejects any set a, b, c which does not satisfy (14).

For example, if $d = -4$, so that $D = 4$, we must have $|b| \leq \sqrt{\tfrac{4}{3}}$ and b even, whence $b = 0$. Now $4ac = 4$, so $a = c = 1$. There is only one reduced form, namely $(1, 0, 1)$. This was the first example used in §5.

To take another case, suppose $d = -7$, so that $D = 7$. Then $|b| \leq \sqrt{\tfrac{7}{3}}$, and b is odd, whence $b = 1$ or -1. Now $4ac = 1 + 7 = 8$, whence $a = 1$, $c = 2$. The possibility that $b = -1$ must be rejected, as it does not comply with (14), and we are left with the single reduced form $(1, 1, 2)$. This was the second example used in §5.

Proceeding in this way, one can easily construct a table of reduced forms. The accompanying Table II covers forms with discriminants from -3 to -83. The forms marked * are the so-called imprimitive forms, that is, forms for which a, b, c have a common factor greater than 1. Such a form is merely a multiple of a primitive form of a previous discriminant.

The reduced forms of a given discriminant constitute a *representative set* of forms of that discriminant, comprising as they do one form out of each class of mutually equivalent forms. The theory of §4 gives the necessary and sufficient condition for a number to be properly representable by one or other of the reduced forms, and this is the result referred to in §1. Where there is only one reduced form, the problem of representation is completely solved. The single reduced form is then the principal form, since the principal form satisfies the conditions for reduction given in (14).

Even where there is more than one reduced form it may be possible to solve the problem of representation. Consider the first such case in the table (excluding imprimitive forms), namely the case $d = -15$. Here there are two reduced forms, $(1, 1, 4)$ and $(2, 1, 2)$. Suppose a number n is represented by the first form; then

$$4n = (2x + y)^2 + 15y^2 \equiv (2x + y)^2 \ (\text{mod } 15).$$

TABLE II

REDUCED POSITIVE DEFINITE FORMS OF
DISCRIMINANT $-D$

D	a, b, c	D	a, b, c	D	a, b, c
3	1, 1, 1	43	1, 1, 11	64	1, 0, 16
4	1, 0, 1	44	1, 0, 11		2, 0, 8*
7	1, 1, 2		2, 2, 6*		4, 0, 4*
8	1, 0, 2		3, 2, 4		4, 4, 5
11	1, 1, 3		3, -2, 4	67	1, 1, 17
12	1, 0, 3	47	1, 1, 12	68	1, 0, 17
	2, 2, 2*		2, 1, 6		2, 2, 9
15	1, 1, 4		2, -1, 6		3, 2, 6
	2, 1, 2		3, 1, 4		3, -2, 6
16	1, 0, 4		3, -1, 4	71	1, 1, 18
	2, 0, 2*	48	1, 0, 12		2, 1, 9
19	1, 1, 5		2, 0, 6*		2, -1, 9
20	1, 0, 5		3, 0, 4		3, 1, 6
	2, 2, 3		4, 4, 4*		3, -1, 6
23	1, 1, 6	51	1, 1, 13		4, 3, 5
	2, 1, 3		3, 3, 5		4, -3, 5
	2, -1, 3	52	1, 0, 13	72	1, 0, 18
24	1, 0, 6		2, 2, 7		2, 0, 9
	2, 0, 3	55	1, 1, 14		3, 0, 6*
27	1, 1, 7		2, 1, 7	75	1, 1, 19
	3, 3, 3*		2, -1, 7		3, 3, 7
28	1, 0, 7		4, 3, 4		5, 5, 5*
	2, 2, 4*	56	1, 0, 14	76	1, 0, 19
31	1, 1, 8		2, 0, 7		2, 2, 10*
	2, 1, 4		3, 2, 5		4, 2, 5
	2, -1, 4		3, -2, 5		4, -2, 5
32	1, 0, 8	59	1, 1, 15	79	1, 1, 20
	2, 0, 4*		3, 1, 5		2, 1, 10
	3, 2, 3		3, -1, 5		2, -1, 10
35	1, 1, 9	60	1, 0, 15		4, 1, 5
	3, 1, 3		3, 0, 5		4, -1, 5
36	1, 0, 9		2, 2, 8*	80	1, 0, 20
	2, 2, 5		4, 2, 4*		2, 0, 10*
	3, 0, 3*	63	1, 1, 16		3, 2, 7
39	1, 1, 10		2, 1, 8		3, -2, 7
	2, 1, 5		2, -1, 8		4, 0, 5
	2, -1, 5		4, 1, 4		4, 4, 6*
	3, 3, 4		3, 3, 6*	83	1, 1, 21
40	1, 0, 10				3, 1, 7
	2, 0, 5				3, -1, 7

Provided n is not divisible by 15, we can easily deduce that n is congruent to one of 1, 4, 6, 9, 10 (mod 15). Similarly, if n is representable by the second form, we find that n is congruent to one of 2, 3, 5, 8, 12 (mod 15). Hence we can distinguish between numbers represented by the one form and numbers represented by the other, except possibly for numbers divisible by 15. The notion of *genus* was introduced by Gauss to express this kind of distinction, and the two forms just considered are said to belong to different genera. But the theory of genera is too extensive and complicated to be developed here, and we must be content with this brief indication.

The possibility we have just discussed, of distinguishing between representation by two different reduced forms, depends on the existence of some modulus (15 in the above example) such that the numbers represented by the two forms satisfy different congruences to that modulus. Where there is no such modulus (and this is indeed the more general case), the problem of representation by an individual form is still essentially unsolved. For example, we can find the condition for a number to be representable by one or other of the forms x^2+55y^2 and $5x^2+11y^2$, but no simple general rule is known for deciding by which of the forms the representation is effected.

8. The number of representations

The theory of §4 gave the necessary and sufficient condition for a number to be properly representable by one or other of the reduced forms of discriminant d; the condition being the solubility of the congruence (8). This theory can be carried a stage further, so as to lead to a determination of the total *number* of proper representations of n by all the reduced forms of discriminant d. We denote this total number by $R(n)$. Where there is only one reduced form of discriminant d (as for instance x^2+y^2 when $d=-4$), the result gives the number of representations by that particular form.

We now outline the theory by which $R(n)$ is determined, but we shall have to pass over the details without proof. For simplicity we shall assume that n is relatively prime to d. This implies, in particular, that any form of discriminant d which represents n is primitive, for if a, b, c had a common factor, this factor would divide both n and d.

The starting point is the same as in §4. We saw there that to each proper representation of n by (a, b, c), say

$$n = ap^2 + bpr + cr^2, \tag{16}$$

there corresponds a substitution which transforms (a, b, c) into an equivalent form (n, h, l) whose first coefficient is n and whose second coefficient satisfies the congruence

$$h^2 \equiv d \,(\mathrm{mod}\ 4n) \tag{17}$$

and the inequality

$$0 \leqq h < 2n. \tag{18}$$

To count the total number $R(n)$ of representations, we have to count how many numbers h satisfy (17) and (18), and then count how many representations such as (16) correspond to the same number h.

Let us begin by considering the latter point. The same number h cannot come from two different reduced forms, for then these forms would both be equivalent to the same form (n, h, l), which is impossible. If two representations of n by (a, b, c) lead to the same number h, then the corresponding substitutions can be combined (by applying first one and then the inverse of the other) so as to give a substitution which transforms (a, b, c) into itself. In fact, it is easily seen that the number of representations of n which give rise to the same number h is equal to the number of unimodular substitutions which transform (a, b, c) into itself.

This brings us to a question not so far considered. A unimodular substitution which transforms a form into itself is called an automorphic substitution, or *automorph*, of the form. There are always two obvious automorphs, namely the identical substitution $x=X$, $y=Y$ and the negative identical substitution $x=-X$, $y=-Y$. In general these are all, but there are two exceptions. The form x^2+y^2 has the two additional automorphs $x=Y$, $y=-X$ and $x=-Y$, $y=X$, making four altogether. The form x^2+xy+y^2 has the four additional automorphs

(i) $x = X + Y, \quad y = -X,$
(ii) $x = -X - Y, \quad y = X,$
(iii) $x = Y, \quad y = -X - Y,$
(iv) $x = -Y, \quad y = X + Y,$

making six altogether. It can be proved that this list of possible automorphs is in fact complete, and the number of automorphs, say w, is therefore 6 if $d=-3$, 4 if $d=-4$, and 2 otherwise. This refers only to primitive forms; the imprimitive form $2x^2+2y^2$ has, of course, the same automorphs as x^2+y^2.

The result is that *the total number R(n) of proper representations of n by all the reduced forms of discriminant d is w times the number of values of h which satisfy the congruence* (17) *and the inequality* (18).

There remains the problem of finding the number of solutions of the congruence (17), and we content overselves here with considering the case $d = -4$. Our previous assumption that n is relatively prime to d means now that n is odd. Cancelling a factor 4 from the congruence (17) and a factor 2 from the inequality (18), we require the number of solutions of

$$h^2 \equiv -1 \pmod{n} \qquad (19)$$

with

$$0 \leq h < n. \qquad (20)$$

By a general principle (II.6), this is the product of the numbers of solutions of the congruences

$$h^2 \equiv -1 \pmod{p^r} \qquad (21)$$

for the various prime powers p^r composing n.

The congruence (21) is insoluble if p is of the form $4k+3$, and has two solutions if p is of the form $4k+1$ and r is 1. By the method used in §5, one can easily prove that in the latter case it still has two solutions if $r>1$. Hence the number of solutions of (19) is 0 if n has any prime factor of the form $4k+3$, and is 2^s if n has s distinct prime factors of the form $4k+1$ and none of the form $4k+3$.

Since $w=4$ for the form x^2+y^2, it follows that *the number of proper representations of an odd number n by the form x^2+y^2 is* 4×2^s *if n has s distinct prime factors of the form $4k+1$ and none of the form $4k+3$*. There are no proper representations if n has any prime factor of the form $4k+3$.

The representations fall into groups of 8, obtained from one another by changing the signs of x and y and interchanging x and y. So the number of essentially different representations, instead of being 4×2^s as above, is 2^{s-1}. This is 1 if n is itself a prime of the form $4k+1$ (as proved in V.2), or if n is a power of such a prime.

9. The class-number

We denote by $C(d)$ the number of classes of forms of discriminant d, that is, the number of reduced forms of discriminant d. For simplicity we shall restrict ourselves to discriminants for which every form is primitive; such discriminants are said to be *fundamental*. A few examples taken from Table II are:

$$C(-3) = 1, \quad C(-4) = 1, \quad C(-51) = 2, \quad C(-71) = 7.$$

We can, of course, interpret $C(d)$ as being the number of sets of integers a, b, c which satisfy $b^2 - 4ac = d$ and also satisfy the inequalities (14) of §6.

A remarkable formula exists for $C(d)$, which makes it possible to determine this number by quite different considerations from any that relate to quadratic forms. The formula is simplest when $d = -p$, where p is a prime, which is necessarily of the form $4k+3$, since $d \equiv 0$ or 1 (mod 4). The case $p = 3$ is, however, exceptional, and we exclude it. We form the sum, say A, of all the quadratic residues (mod p), and the sum B of all the quadratic non-residues. Then

$$C(-p) = \frac{B - A}{p}. \tag{22}$$

For example, if $p = 23$, the quadratic residues are

$$1, 2, 3, 4, 6, 8, 9, 12, 13, 16, 18, \text{ with sum } 92,$$

and the quadratic non-residues are

$$5, 7, 10, 11, 14, 15, 17, 19, 20, 21, 22, \text{ with sum } 161.$$

The formula gives

$$C(-23) = \frac{161 - 92}{23} = 3,$$

which is correct, as one sees from the table.

The honour of having discovered this remarkable formula seems to rest with Jacobi, though the discovery may also have been made independently by Gauss. Jacobi proved that the number $\dfrac{B - A}{p}$ has a certain property in common with the class-number $C(-p)$, and then by examining many numerical instances he came to the conclusion that the two were no doubt always equal. This he announced in 1832, but confessed himself unable to give a proof. The first published proof was that given by Dirichlet in 1838, and the formula is generally called Dirichlet's class-number formula. Dirichlet's proof used infinite series, and was intimately connected with his proof of the existence of primes in arithmetical progressions. Ever since Dirichlet's proof, mathematicians have sought an elementary proof of the class number formula, i.e. a proof that does not involve a limit process. Finally, in 1978, M. L. S. Orde gave such a proof for the case of negative discriminants.

The fact that $B-A$ is a multiple of p, and indeed that A and B are both multiples of p, is quite elementary. The quadratic residues are congruent to 1^2, 2^2, ... , $(\frac{1}{2}(p-1))^2$, and it is easy to evaluate this sum and see that it is a multiple of p. So A is a multiple of p, and since $A+B=1+2+...+(p-1)=\frac{1}{2}(p-1)p$, it follows that B is also a multiple of p.

There are several other formulae for $C(-p)$ which are equivalent to (22), and some of them are more convenient for numerical work. We have selected this particular one because it is easy to formulate, and does not require any division into cases, as some of the others do. The various formulae can all be extended to the case when d is not necessarily of the form $-p$.

As regards the magnitude of the class-number, Gauss conjectured from extensive numerical evidence that $C(d)$ tends to infinity as d tends to $-\infty$. This conjecture was first proved by Heilbronn in 1934, and his proof represented an important step forward in analytic number-theory.

It has long been known that $C(d)=1$ when $-d$ has the nine values:

$$3, 4, 7, 8, 11, 19, 43, 67 \text{ and } 163.$$

Heilbronn and Linfoot proved in 1934 that there is at most one more negative discriminant with this property. Numerical evidence suggested that in fact there was no such 'tenth discriminant', but the question was not settled to everyone's satisfaction until 1966, when a complete proof was given by H. Stark. Another method of proof was found at about the same time by A. Baker. Proofs were also found by Deuring and Siegel. Some time later, a proof given by K. Heegner in 1952, the validity of which had been questioned, was accepted as indeed being a valid proof.

NOTES

1. There are two notations in common use for the general quadratic form. One is that which we have adopted: $ax^2+bxy+cy^2$. The other is $ax^2+2bxy+cy^2$, which presupposes that the middle coefficient is even. The latter notation excludes such a form as x^2+xy+y^2, though of course its properties can be deduced from those of $2x^2+2xy+2y^2$,

which is admitted. The notation without the factor 2 was used by Lagrange, Kronecker and Dedekind, that with the factor 2 was used by Legendre, Gauss and Dirichlet. As one might expect from seeing these great names on both sides, neither notation has a decisive superiority over the other. The position is that some results take a simpler form when the first notation is used and others take a simpler form when the second is used.

The most accessible accounts of the theory available in English are those given in Mathews's *Theory of Numbers* and in Dickson's *Introduction to the Theory of Numbers* or *Modern Elementary Theory of Numbers*. Dickson uses the Lagrange notation, as we have done, and Mathews uses the Gauss notation. We must refer the reader to Dickson's *Introduction* for proofs of the various results which are stated without proof in the present chapter. For an account of the general theory of quadratic forms (not only binary forms), see B. W. Jones, *The Arithmetic Theory of Quadratic Forms* (Carus Monograph no. 10, 1950), G. L. Watson, *Integral Quadratic Forms* (Cambridge Tracts, no. 51, 1960) or O. T. O'Meara, *Introduction to Quadratic Forms* (Springer, 1963). For an interesting account of the theory of general quadratic forms over the rational field, see J. W. S. Cassels, *Rational Quadratic Forms* (Academic Press, London, 1978).

§2. Forms which are related by a substitution of determinant -1 are said to be improperly equivalent. The use of substitutions of determinant -1 complicates the theory of automorphs, both for definite and indefinite forms.

§8. From the number of *proper* representations of a number as the sum of two squares, found in this section, one can deduce the formula $4(D_1 - D_3)$, mentioned in the Notes on Chapter V, for the total number of representations (proper and improper), and in this formula it is not necessary that n should be odd.

§9. For Jacobi's investigation, see Bachmann, *Die Lehre von der Kreisteilung* (Teubner, 1927), p. 292. For a proof of Dirichlet's class-number formula, see Landau's *Vorlesungen*, vol. I, pp. 127–80, or Mathews, ch. 8. The latter exposition uses the Gauss notation, and therefore the formula is a little different.

Orde's elementary proof can be found in *J. London Math. Soc.*, (2) 18 (1978), 409–20.

For the work of Heilbronn, and of Heilbronn and Linfoot, see *Quart. J. of Math.*, 5 (1934), 150–60 and 293–301. Stark's paper is in *Michigan Math. J.*, 14 (1967), 1–27. For Baker's method see *Mathematika*, 13 (1966), 204–16 (205). For Deuring's proof see *Inventiones Math.*, 5 (1968), 169–79. For Siegel's proof see *ibid.*, 180–91. For Heegner's proof see *Mathematische Zeitschrift*, 56 (1952), 227–53 and *J. Number Theory*, 1 (1969), 16–27.

VII

SOME DIOPHANTINE EQUATIONS

1. *Introduction*

A Diophantine equation, or indeterminate equation, is one which is to be solved with integral values for the unknowns. We have already met some classical Diophantine equations, for example the equation $x^2+y^2=n$ in Chapters V and VI, and the equation $x^2-Ny^2=1$ in Chapter IV.

There is probably no branch of the theory of numbers which presents greater difficulties than the theory (if it can be called a theory) of Diophantine equations. A glance at the extensive literature gives one an impression of a mass of unrelated results on miscellaneous special equations, discovered by highly ingenious devices, which do not seem to fit together into any general theory. After an equation has been solved by some special device, a theory has sometimes been constructed round the solution, which exhibits it in a more reasonable light and enables one to see how far it can be generalized. But the intrinsic difficulties of the subject are so great that the scope of any such theory is usually very limited. Where an extensive theory has developed out of Diophantine equations of a particular type, as with the theory of quadratic forms, it has soon been regarded as having attained an independent status.

In this chapter we shall discuss some Diophantine equations which admit of elementary treatment, and shall mention where possible any general theories which may be associated with them.

2. The equation $x^2+y^2=z^2$

Numerical solutions of this equation, such as $3^2+4^2=5^2$, have been known from an early period in man's history. A Babylonian tablet has survived, dating from about 1700 B.C., which contains what is in effect an extensive list of solutions, some of the numbers being quite large. The equation was naturally of great interest to the Greek mathematicians, because of its connection with the theorem of Pythagoras, and the general solution is given in Euclid (Book X, Lemma 1 to Prop. 29).

If we divide the equation throughout by z^2, and put $\frac{x}{z}=X$, $\frac{y}{z}=Y$, it becomes

$$X^2 + Y^2 = 1, \qquad (1)$$

and the problem is reduced to that of finding the solutions of this equation in rational numbers X, Y. The appropriate treatment of the equation is suggested by writing it as

$$Y^2 = 1 - X^2 = (1 - X)(1 + X).$$

We cannot express X rationally in terms of $(1-X)(1+X)$, but we can express it rationally in terms of $(1-X)/(1+X)$. We therefore divide throughout by $(1+X)^2$, getting

$$\left(\frac{Y}{1 + X}\right)^2 = \frac{1 - X}{1 + X}.$$

If we put $t=Y/(1+X)$, then both X and Y are expressible as rational functions of t; we have

$$\frac{1 - X}{1 + X} = t^2,$$

whence

$$X = \frac{1 - t^2}{1 + t^2}, \quad Y = \frac{2t}{1 + t^2}. \qquad (2)$$

For every rational number t, these formulae give rational numbers X, Y which satisfy (1). Conversely, every rational solution of (1) is obtained in this way, apart from the special solution $X=-1$, $Y=0$, which is approached if t is taken arbitrarily large but is not itself representable in the form (2).

The preceding argument can also be looked at from a geometrical point of view. The equation $X^2+Y^2=1$ represents a circle, with centre at the origin of coordinates and radius 1. Take a particular point on the circle, say the point $X=-1$, $Y=0$. A variable line drawn through this point will meet the circle in one other point (except when it happens to be a tangent), and the coordinates of this other point can be found from the equations of the circle and the straight line by rational operations. A variable line through the point $(-1, 0)$ has an equation of the form $Y=t(X+1)$, and the formulae (2) express the coordinates of the point of intersection in terms of t. A similar method can be used to find the rational points on a conic, provided that the equation to the conic has rational coefficients, and provided that we can find some one rational point on the curve. This, however, may not be possible; for example there is no rational point on the circle $X^2+Y^2=3$. Or, even if there are rational points on a conic, it may not be an easy matter to find one.

The formulae (2), where t is any rational number, give the general solution of the equation $X^2+Y^2=1$ in rational numbers, and therefore in principle they give the general solution of the equation

$$x^2 + y^2 = z^2 \qquad (3)$$

in integers. But the transition from the rational solutions of (1) to the integral solutions of (3) raises a question which calls for consideration, and sometimes in other problems presents serious difficulties. Put $t=\dfrac{q}{p}$, where p and q are relatively prime integers. Then, by (2),

$$\frac{x}{z} = \frac{p^2 - q^2}{p^2 + q^2}, \quad \frac{y}{z} = \frac{2pq}{p^2 + q^2} . \qquad (4)$$

It is certainly *possible* to take x, y, z to be p^2-q^2, $2pq$, p^2+q^2, or to be any common multiple of these numbers, and we shall then have a solution in integers of the equation (3). But it is not certain that x, y, z *must* be common multiples of these numbers. If the three numbers p^2-q^2, $2pq$, p^2+q^2 have a common factor greater than 1, we can divide them by this common factor and still get a solution of (3) in integers.

We consider two possibilities for the relatively prime integers p and q. First suppose that one of them is even and the other odd. Then the three numbers p^2-q^2, $2pq$, p^2+q^2 have no common factor greater than 1, for such a factor would have to be odd (since p^2-q^2 is odd) and would have to divide $(p^2-q^2)+(p^2+q^2)=2p^2$, and similarly would have to divide $2q^2$, and this is impossible since p and q are relatively prime. Hence, in this case, it follows from (4) that

$$x = m(p^2 - q^2), \quad y = 2mpq, \quad z = m(p^2 + q^2), \qquad (5)$$

where m is an integer.

Next consider the possibility that p and q are both odd. In this case, if we put $p+q=2P$ and $p-q=2Q$, the numbers P and Q are relatively prime integers. One of them is even and one odd, since $P+Q=p$ is odd. Substituting for p and q in terms of P and Q in (4), we obtain

$$\frac{x}{z} = \frac{2PQ}{P^2 + Q^2}, \quad \frac{y}{z} = \frac{P^2 - Q^2}{P^2 + Q^2},$$

after cancelling a factor 2. The position is therefore the same as before, except that x and y are interchanged, and P and Q take the place of p and q.

It follows that *all solutions of $x^2+y^2=z^2$ in integers are given by the formulae (5), where m, p, q are integers, and p and q are relatively prime, and one of them is even and the other odd*, apart from the possibility of interchanging x and y. These are the formulae of Euclid. The simplest solution (apart from trivial solutions with one of the unknowns zero) is $x=3$, $y=4$, $z=5$, which arises by putting $m=1$, $p=2$, $q=1$. The first few primitive solutions (that is, solutions with x, y, z relatively prime, and therefore $m=1$) are $(3, 4, 5)$, $(5, 12, 13)$, $(8, 15, 17)$, $(7, 24, 25)$, $(21, 20, 29)$, $(9, 40, 41)$.

Since the formula for z (taking m to be 1) is $z=p^2+q^2$, we can make z a perfect square by choosing p and q suitably, and so obtain a parametric solution for $x^2+y^2=z^4$. Repetition of the process enables one to give solutions for $x^2+y^2=z^k$, where k is any power of 2. Alternatively, the formulae for such an equation could be deduced from the formulae for $x^2+y^2=z^2$ by employing the identity (1) of Chapter V.

3. The equation $ax^2+by^2=z^2$

The method used above for the equation $x^2+y^2=z^2$ would also apply to the equation $ax^2+y^2=z^2$, and would again lead to formulae for the general solution. As before, there are infinitely many primitive solutions. But the method will *not* apply to the more general equation

$$ax^2 + by^2 = z^2, \qquad (6)$$

where a and b are natural numbers, neither of which is a perfect square.

Indeed, a moment's consideration shows that such an equation may not be soluble (apart from the solution $x=y=z=0$, which we shall exclude throughout). For example, the equation

$$2x^2 + 3y^2 = z^2$$

is insoluble. For we can suppose that x, y, z have no common factor greater than 1, whence it follows in particular that neither x nor z is divisible by 3. But then the congruence $2x^2 \equiv z^2$ (mod 3) is impossible, since 2 is a quadratic non-residue to the modulus 3.

Similar considerations apply to the general equation (6), and give congruence conditions which must be satisfied if the equation is to be soluble. We can suppose that a and b are both *square free*, that is, not divisible by any square greater than 1; for the introduction of square factors into the coefficients a and b does not affect the solubility of the equation.

If the equation (6) is soluble, we can divide out any common factor of x, y, z and so obtain a solution in which x, y, z have no common factor greater than 1. The equation implies the congruence $ax^2 \equiv z^2$ (mod b). Now x and b must be relatively prime; for if they had a prime factor in common, this prime would divide x and z, and therefore its square would divide by^2, and since b is square free this would require the prime to divide y, which is impossible. Multiplying the congruence throughout by x'^2, where $xx' \equiv 1$ (mod b), we obtain a congruence of the form

$$a \equiv \alpha^2 \text{ (mod } b\text{)}, \tag{7}$$

where $\alpha = x'z$. Similarly

$$b \equiv \beta^2 \text{ (mod } a\text{)} \tag{8}$$

for some integer β. That is, a must be a quadratic residue (mod b), and b must be a quadratic residue (mod a). Here we are using the term *quadratic residue* in a more general sense than in Chapter III, since the moduli a and b are now not necessarily primes.

If a and b have H.C.F. $h>1$, there is another congruence besides (7) and (8) which must be soluble if the equation (6) is to be soluble. Put $a=ha_1$ and $b=hb_1$, so that a_1, b_1, h are relatively prime in pairs. In any solution of (6), z must be divisible by h, so that

$a_1x^2+b_1y^2$ must be divisible by h. Multiplying throughout by $b_1x'^2$, we obtain a congruence of the form

$$a_1b_1 \equiv - \gamma^2 \text{ (mod } h). \qquad (9)$$

The fact that the congruences (7), (8), (9) must be soluble imposes restrictions on a and b which are necessary for the solubility of the equation (6). It is by no means obvious that if the congruences are soluble then the equation is soluble. We shall now prove, following Legendre, that this is in fact the case, and so shall establish that *the equation* (6), *where a and b are square free natural numbers, is soluble if and only if the congruences* (7), (8), (9) *are all soluble.*

If either a or b is 1, the equation is obviously soluble. If $a=b$, the congruence conditions (7) and (8) are trivially satisfied, and (9) reduces to $1 \equiv -\gamma^2$ (mod a). By VI.5, this implies that a is representable as p^2+q^2, and the equation is satisfied by $x=p$, $y=q$, $z=p^2+q^2$.

We can now suppose that $a>b>1$. The plan of the proof is to derive from (6) a similar equation with the same b but with a replaced by A, where $0<A<a$, and A, b satisfy the same three congruence conditions as a, b. Repetition of the process must lead eventually to an equation in which either one coefficient is 1 or the two coefficients are equal. As we have seen, such an equation is soluble.

By hypothesis, the congruence (8) is soluble. We choose a solution β which satisfies $|\beta| \leqq \frac{1}{2}a$. Since β^2-b is a multiple of a, we can put

$$\beta^2 - b = aAk^2, \qquad (10)$$

where k and A are integers and A is square free (all the square factors being absorbed in k^2). We note that k is relatively prime to b, since b is square free. We observe that A is positive, since

$$aAk^2 = \beta^2 - b > - b > - a,$$

whence $Ak^2 \geqq 0$, and therefore >0 since b is not a perfect square.

If we substitute for y and z in terms of new variables Y and Z from*

$$z = bY + \beta Z, \quad y = \beta Y + Z, \qquad (11)$$

*The form of the substitution (11) is suggested by writing
$$z-y\sqrt{b} = (\beta-\sqrt{b})(Z - Y\sqrt{b}).$$

we find that

$$z^2 - by^2 = (\beta^2 - b)(Z^2 - bY^2).$$

In view of (10), the equation (6) becomes

$$ax^2 = aAk^2(Z^2 - bY^2).$$

Putting $x=kAX$, the new equation becomes

$$AX^2 + bY^2 = Z^2.$$

If this equation is soluble, so is (6); for the substitution (11) and the equation $x=kAX$ give integral values, not all zero, for x, y, z in terms of X, Y, Z.

The new coefficient A is positive and square free, and satisfies

$$A = \frac{1}{ak^2}(\beta^2 - b) < \frac{\beta^2}{ak^2} \leqq \frac{\beta^2}{a} \leqq \frac{1}{4}a,$$

and therefore A is less than a. It remains to be proved that A and b satisfy the congruence conditions analogous to (7), (8), (9). The analogue of (8) is obvious, since $b \equiv \beta^2 \pmod{A}$ by (10).

To prove the analogue of (7), we observe that (10) can be divided throughout by h, giving

$$h\beta_1{}^2 - b_1 = a_1 Ak^2.$$

Also (7) is equivalent to $a_1 \equiv h\alpha_1{}^2 \pmod{b_1}$. Hence

$$h\beta_1{}^2 \equiv hA(\alpha_1 k)^2 \pmod{b_1},$$

and since h, k, a_1 are all relatively prime to b_1 it follows that A is congruent to a square $\pmod{b_1}$. Also $-a_1 Ak^2 \equiv b_1 \pmod{h}$, and in view of (9) and the fact that k, a_1, b_1 are all relatively prime to h it follows that A is congruent to a square $\pmod{h}$, and therefore also $\pmod{b}$, giving the analogue of (7).

To prove the analogue of (9) with A in place of a, let H denote the highest common factor of A and b, and put $A=HA_2$, $b=Hb_2$. The equation (10) can be divided by H, giving

$$H\beta_2{}^2 - b_2 = aA_2 k^2.$$

Hence

$$-A_2 b_2 \equiv a(A_2 k)^2 \pmod{H}.$$

Since $a \equiv \alpha^2 \pmod{H}$ by (7), it follows that $-A_2 b_2$ is congruent to a square $\pmod{H}$, which is the analogue of (9).

We have now shown that the coefficients A and b satisfy similar congruence conditions to those imposed on a and b. The method of proof already explained therefore applies, and establishes the solubility of the equation (6).

To illustrate the above proof, we apply the process to the equation

$$41x^2 + 31y^2 = z^2. \tag{12}$$

Since the coefficients are relatively prime, there are only the two congruence conditions

$$41 \equiv \alpha^2 \,(\text{mod } 31) \quad \text{and} \quad 31 \equiv \beta^2 \,(\text{mod } 41).$$

These are both soluble, namely with $\alpha \equiv \pm 14$ (mod 31), and $\beta \equiv \pm 20$ (mod 41). Indeed, in this particular case, the solubility of one congruence implies that of the other by the law of quadratic reciprocity (III.5), since 31 and 41 are primes and are not both of the form $4k+3$.

To follow the method, we must choose a value for β and then define A and k by (10). In the theory, we supposed $|\beta| \leqq \frac{1}{2}a$, so we take $\beta = 20$, and have $\beta^2 - b = 400 - 31 = 9 \times 41$, hence $k = 3$ and $A = 1$. (The fact that $A = 1$ means that no further repetition of the process will be necessary.) The new equation derived from (12) is $X^2 + 31Y^2 = Z^2$, and we take the obvious solution $X = 1$, $Y = 0$, $Z = 1$. The relations between x, y, z and X, Y, Z with the coefficients now in use are

$$z = 31Y + 20Z, \quad y = 20Y + Z, \quad x = 3X.$$

These give the solution $x = 3$, $y = 1$, $z = 20$ for the original equation (12).

We now return to the general theory. We have proved that the solubility of the congruences (7), (8), (9) is necessary and sufficient for the solubility of the equation (6), on the supposition that a and b are square free. Legendre easily deduced from this result a necessary and sufficient condition for the solubility of the equation

$$ax^2 + by^2 = cz^2,$$

where a, b, c are natural numbers. On the supposition that a, b, c are square free and relatively prime in pairs (which are not serious restrictions here), the condition is that the three congruences

$$bc \equiv \alpha^2 \,(\text{mod } a), \quad ca \equiv \beta^2 \,(\text{mod } b), \quad ab \equiv -\gamma^2 \,(\text{mod } c)$$

must all be soluble.

We conclude this section with some remarks on the general question of congruence conditions for the solubility of Diophantine equations. Any Diophantine equation gives rise to a congruence to any modulus we care to select, and every such congruence must be soluble if the equation is to be soluble. Usually there are only a finite number of moduli for which the solubility of the congruence imposes any conditions on the coefficients of the equation. The resulting conditions are *necessary* conditions for the equation to be soluble. They are not always sufficient, and the elucidation of the relation between the solubility of congruences and of equations raises deep and delicate questions. As we have said, the congruence conditions are both necessary and sufficient for the solubility of Legendre's equation $ax^2+by^2=cz^2$. It was proved by Hasse in 1923 that a similar result holds for homogeneous quadratic equations in any number of variables.

We have already met various instances in which an equation is proved to be insoluble by congruence considerations. It is sometimes possible to prove the insolubility of an equation by using a congruence to a modulus *which depends on the unknowns in the equation*. This is the underlying idea of the proof, given by V. A. Lebesgue in 1869, that the equation

$$y^2 = x^3 + 7$$

is insoluble in integers. First, x must be odd, since a number of the form $8k+7$ cannot be a square. Now write the equation as

$$y^2 + 1 = x^3 + 8 = (x + 2)(x^2 - 2x + 4).$$

The number $x^2-2x+4=(x-1)^2+3$ is of the form $4k+3$. Hence it has some prime factor q of that form, and since the congruence $y^2+1\equiv0 \pmod{q}$ is insoluble, the proposed equation is insoluble.

4. *Fermat's problem*

Much of our knowledge of Fermat's discoveries is derived from the comments which he wrote on the margin of his copy of the *Arithmetic* of Diophantus. Opposite the account of the equation $x^2+y^2=z^2$ in Diophantus, Fermat wrote: 'However, it is impossible to write a cube as the sum of two cubes, a fourth power as the sum of two fourth powers, and in general any power beyond the second as the sum of two similar powers. For this I have discovered a truly wonderful proof, but the margin is too small to contain it.' This is the famous conjecture of Fermat, generally called Fermat's Last Theorem, namely that the equation

$$x^n + y^n = z^n \tag{13}$$

has no solution in natural numbers x, y, z if n is an integer greater than 2. Despite the efforts of many of the greatest mathematicians of the last 300 years, it remains unproved as a general proposition, though it has been proved to be true for every value of n up to about 125,000. Most probably Fermat was mistaken in thinking that he had a proof.

The attraction of the problem lies partly in the tantalizing simplicity of its formulation. For this reason it has obsessed many amateurs whose self-confidence has been greater than their mathematical ability, and it certainly has the distinction of being the arithmetical problem for which the greatest number of incorrect 'proofs' has been put forward.

It seems likely that any new method devised for the proof of Fermat's conjecture would lead to important new developments in the theory of numbers generally. This was indeed amply realized in the case of the work of Kummer (1810–93). Kummer believed at first that he had proved Fermat's conjecture. The fallacy in his arguments was pointed out to him by Dirichlet, and Kummer's efforts to repair the mistake led him to create a new and extensive theory, that of *ideals* in algebraic number-fields.

In an elementary account such as this, we must content ourselves with proving the truth of Fermat's conjecture for some particular value of n. The simplest case to treat is $n=4$, where the insolubility of the equation was proved by Fermat himself.

Fermat proved, more generally, that the equation

$$x^4 + y^4 = z^2 \tag{14}$$

has no solution in natural numbers, and his proof is an outstanding example of his technique of 'infinite descent', which is simply another form of the principle of proof by induction. From any one hypothetical solution of the equation in natural numbers, Fermat derived another with a smaller value of z. Repetition of the process leads eventually to a contradiction, since a decreasing sequence of natural numbers cannot continue indefinitely. The principle is the same as that underlying Legendre's method, described in the previous section, except that here it is used to prove insolubility, whereas there it was used to prove solubility.

Suppose x, y, z are natural numbers which satisfy (14). We can suppose that x and y have no common factor greater than 1, for the fourth power of such a common factor can be cancelled from the equation. The numbers x^2, y^2, z constitute a primitive solution of $X^2 + Y^2 = Z^2$, and therefore, by the result proved in §2, they are expressible (possibly after interchanging x and y) as

$$x^2 = p^2 - q^2, \quad y^2 = 2pq, \quad z = p^2 + q^2,$$

where p and q are relatively prime natural numbers, one of which is even and the other odd. Looking at the first equation, and recalling that any square must be congruent to 0 or 1 (mod 4), we see that p must be odd and q even. Putting $q=2r$, we have

$$x^2 = p^2 - (2r)^2, \quad (\tfrac{1}{2}y)^2 = pr.$$

Since p and r are relatively prime and their product is a perfect square, each of them must be a perfect square. If we put $p=v^2$ and $r=w^2$, the first equation becomes

$$x^2 + (2w^2)^2 = v^4.$$

This equation is somewhat similar to (14) in its general form. When similar reasoning is applied again to the new equation, we obtain one exactly like (14). The last equation implies that

$$x = P^2 - Q^2, \quad 2w^2 = 2PQ, \quad v^2 = P^2 + Q^2,$$

where P and Q are relatively prime natural numbers, one of which is even and the other odd. Since $PQ=w^2$, each of P and Q must be a perfect square. Putting $P=X^2$, $Q=Y^2$, the third equation becomes

$$X^4 + Y^4 = v^2,$$

which is of the same form as (14). In this equation X, Y, v are natural numbers, and

$$v^2 = p < \sqrt{z},$$

whence $v<z$. In view of what was said earlier, this is enough to prove the insolubility of the equation (14).

Modern researches on Fermat's problem have almost all been based on the work of Kummer. They have resulted in proofs that if n satisfies any one of a series of conditions then the equation (13) is insoluble. It is enough to consider prime values of n greater than 2, because any number greater than 2 is either divisible by some prime greater than 2, or else divisible by 4; and if the equation is insoluble for one value of n it is *a fortiori* insoluble for any multiple of that value. So far, whenever a number n has been reached which did not satisfy any of the existing criteria, it has generally been possible to find another criterion which would cope with the number.

5. The equation $x^3+y^3=z^3+w^3$

Although the equation $x^3+y^3=z^3$ (a special case of Fermat's equation) is insoluble, the equation $x^3+y^3=z^3+w^3$ has infinitely many solutions in integers, other than the obvious solutions with $x=z$ or $x=w$ or $x=-y$. Formulae giving solutions were found by Vieta in 1591, but the formulae discovered by Euler in 1756–60 are more general. These were simplified by Binet in 1841.

To treat the equation

$$x^3 + y^3 = z^3 + w^3, \tag{15}$$

put $x+y=X$, $x-y=Y$, $z+w=Z$, $z-w=W$. The equation becomes

$$X(X^2 + 3Y^2) = Z(Z^2 + 3W^2). \tag{16}$$

There is an identity, similar to (1) of Chapter V, which expresses the product of two numbers of the form X^2+3Y^2 as itself of that form, namely

$$(X^2 + 3Y^2)(Z^2 + 3W^2) = (XZ + 3YW)^2 + 3(YZ - XW)^2.$$

If we multiply (16) throughout by X^2+3Y^2, and divide by Z, the identity gives

$$\frac{X}{Z}(X^2 + 3Y^2)^2 = (XZ + 3YW)^2 + 3(YZ - XW)^2.$$

This shows that the rational number $\dfrac{X}{Z}$ is of the form p^2+3q^2, where p and q are the rational numbers given by

$$p = \frac{XZ + 3YW}{X^2 + 3Y^2}, \quad q = \frac{YZ - XW}{X^2 + 3Y^2}. \tag{17}$$

To simplify the algebra, we put $Z=1$ and consider X, Y, W as rational numbers. By (17), with $Z=1$, we have

$$pX + 3qY = 1, \quad pY - qX = W.$$

These formulae allow one to express Y and W in terms of p, q and X, where $X=p^2+3q^2$. They give

$$3qY = 1 - pX, \quad 3qW = p - X^2.$$

If we go back to the original x, y, z, w and remove the obvious denominator, we obtain

$$\begin{cases} x = 1-(p-3q)(p^2+3q^2), & y = -1+(p+3q)(p^2+3q^2), \\ z = p+3q-(p^2+3q^2)^2, & w = -(p-3q)+(p^2+3q^2)^2. \end{cases} \quad (18)$$

These are the formulae of Euler and Binet. For any rational numbers p and q, they give rational numbers x, y, z, w which satisfy the equation (15), and the proof shows that conversely every rational solution of (15) is proportional to a solution provided by these formulae.

If in particular we give p and q integral values, we obtain integral solutions of (15), but there is no reason to expect that every integral solution will be obtainable in this way. One particular solution, obtained by putting $p=1$, $q=1$ is $x=9$, $y=15$, $z=-12$, $w=18$, corresponding to the curious fact that $3^3+4^3+5^3=6^3$. The values $p=4$, $q=1$ correspond to

$$3^3 + 60^3 = 22^3 + 59^3.$$

The simplest solution of (15) with x, y, z, w all positive is

$$1^3 + 12^3 = 9^3 + 10^3.$$

The number 1729 is in fact the least number which is expressible as the sum of two positive integral cubes in two different ways.*

An interesting identity, to which Mahler drew attention in 1936, is obtained by putting $p=3q$. This gives $x=1$, $y=-1+72q^3$, $z=6q-144q^4$, $w=144q^4$. Writing $2q=t$, we obtain the identity

$$(1 - 9t^3)^3 + (3t - 9t^4)^3 + (9t^4)^3 = 1.$$

The interest of this lies in the fact that it shows that the number 1 can be represented in an infinity of ways as the sum of three integral cubes. There is a similar identity for the number 2. I do not know of any identity which exhibits the number 3 as a sum of three integral cubes in infinitely many ways.

*When Hardy visited Ramanujan, who was lying ill at Putney, he mentioned that he had come in taxi no. 1729, and that the number seemed to him rather a dull one, whereupon Ramanujan immediately recalled this special property of the number.

It may be appropriate to mention at this point another unsolved problem. Not every number can be represented as the sum of three integral cubes; indeed, no number congruent to 4 or 5 (mod 9) can be so represented. For it is easily verified that any cube is congruent to 0 or 1 or -1 to the modulus 9, and consequently the sum of any three integral cubes must be congruent to 0 or ± 1 or ± 2 or ± 3 (mod 9), and can never be congruent to ± 4. The problem is: *is every number representable as the sum of four integral cubes?* Despite many attempts, this is still unsolved.

There is a very simple way of expressing any number as the sum of *five* integral cubes. We have

$$(x + 1)^3 + (x - 1)^3 + (-x)^3 + (-x)^3 = 6x.$$

Hence any multiple of 6 is representable by four integral cubes. Now any number can be reduced to a multiple of 6 by subtracting a suitable cube. Indeed, it is easily seen that $n - n^3$ is always a multiple of 6. This gives the result, which seems to have been first proved by Oltramare in 1894.

6. *Further developments*

Many modern researches on Diophantine equations are based on a method originated by the Norwegian mathematician Axel Thue in 1908. This method depends on consideration of the rational approximations to an algebraic number, and a few words of explanation are therefore necessary.

Suppose $f(x, y)$ is any homogeneous form in x and y of degree n, say

$$f(x, y) = a_0 x^n + a_1 x^{n-1} y + \dots + a_n y^n,$$

where $a_0, a_1, \dots, a_n$ are integers, and n is at least 3. We suppose that the form is irreducible, that is, cannot be expressed as the product of two other forms with rational coefficients.* By the so-called fundamental theorem of algebra, the form can be factorized as

$$a_0(x - \theta_1 y)(x - \theta_2 y) \dots (x - \theta_n y),$$

where $\theta_1, \theta_2, \dots, \theta_n$ are irrational numbers, real or complex. These numbers are the roots of the irreducible algebraic equation

---

*Whether we say *rational* coefficients or *integral* coefficients makes no difference, as it can be proved that a factorization into forms with rational coefficients implies a factorization into forms with integral coefficients.

$$a_0\theta^n + a_1\theta^{n-1} + \ldots + a_n = 0$$

and are said to be *algebraic numbers* of degree n.

Whatever integral values we give to x and y, the value of $f(x, y)$ is an integer. Hence, if x and y are not both zero, we have

$$| a_0 (x - \theta_1 y)(x - \theta_2 y) \ldots (x - \theta_n y) | \geqq 1.$$

Now suppose that $\dfrac{x}{y}$ is a rational approximation to θ_1, with y a large positive integer. Then all the factors $x - \theta_2 y, \ldots$ are less than some constant multiple of y, and it follows on division by y^n that

$$\left| \frac{x}{y} - \theta_1 \right| > \frac{K}{y^n}, \tag{19}$$

where K is a positive constant, depending on the particular form f. Thus an algebraic number of degree n cannot have a sequence of rational approximations which approach it too rapidly. The result was found by Liouville in 1844, and was used by him to construct numbers which are not algebraic.

Thue proved, by a long and difficult train of reasoning, that a substantially better inequality is true, namely that

$$\left| \frac{x}{y} - \theta_1 \right| > \frac{1}{y^\nu} \tag{20}$$

for all but a finite number of rational approximations to θ_1, where ν is any number greater than $\tfrac{1}{2}n+1$. The number $\tfrac{1}{2}n+1$ was substantially reduced by Siegel in 1921 to a little less than $2\sqrt{n}$ and further by Dyson and independently by Gelfond to $\sqrt{(2n)}$ in 1947.

In 1955 Roth proved the remarkable theorem that if ν is any number greater than 2, the inequality (20) holds for all but a finite number of rational approximations to θ_1. This is the best possible result of its kind, for as we have seen in IV.7, the inequality

$$\left| \frac{x}{y} - \theta_1 \right| < \frac{1}{y^2}$$

always has infinitely many solutions, whether θ_1 is an algebraic number or not, provided that it is irrational. The proof of Roth's theorem is naturally very difficult.

The inequality (20) leads to a lower bound for the form $f(x, y)$. If x, y are any large integers for which $| f(x, y) |$ is small compared with $|y|^n$, then $\dfrac{x}{y}$ must be a rational approximation to one of the

roots $\theta_1, \ldots, \theta_n$. Supposing, as we may without loss of generality,

that $\dfrac{x}{y}$ is an approximation to θ_1, it follows from (20) that

$$|f(x, y)| > K_1 y^{n-v},$$

where K_1 is some positive constant. We can take v to be any number greater than 2, by Roth's result. Hence any Diophantine equation which implies that $|f(x, y)|$ is *less* than a certain power of $|y|$ can have only a finite number of solutions. In particular, an equation of the form

$$f(x, y) = g(x, y),$$

where $g(x, y)$ is any polynomial, homogeneous or not, in which every term is of degree less than $n-2$, can have only a finite number of solutions. As a special case, this holds if $g(x, y)$ is a constant. It is essential, of course, that n should be at least 3. As we know, Pell's equation $x^2 - Ny^2 = 1$, of degree 2, has infinitely many solutions.

As an illustration, we may consider any equation of the form

$$ax^4 + bx^3y + cx^2y^2 + dxy^3 + ey^4 = kx + ly + m.$$

This has only a finite number of solutions, provided that the form on the left is irreducible. For the right-hand side is of degree 1, and $1 < n-2$ when $n=4$.

The Thue–Siegel–Roth method has one peculiar feature. Although it proves that various types of equation in two variables x and y have only a finite number of solutions, it does not seem to give any limits for x and y beyond which there is no solution. The reason for this failure is that the method is based on the consideration of *two or more* hypothetical approximations to an algebraic number. A contradiction is obtained if all of them are 'too good'. Hence it is generally possible, in any particular case, to deduce limits for x and y beyond which the equation has *at most one* solution, or at most a specified number of solutions, but not limits beyond which the equation has no solution.

This is a serious limitation on the value of the Thue–Siegel–Roth theorem, from the point of view of finding all the solutions of a particular Diophantine equation. We can get an estimate for their number (for the types of equation discussed above), but unless by extreme good fortune we actually find this number of solutions, we cannot be sure, however far we go in searching for a solution, that there are no more.

Recent work by A. Baker has added greatly to our knowledge in this respect. He has found limits for all the solutions of Diophantine equations of certain classes; these classes, though less extensive than those to which the Thue–Siegel–Roth theorem applies, include all equations of the type

$$f(x, y) = m,$$

where f is an irreducible form of degree 3 or more. An explicit bound is established for $|x|$ and $|y|$ in terms of m and the coefficients of f. Thus it becomes possible to find all the solutions of any particular equation of this type by a limited number of trials (though the number may be large). The same applies to equations of the type

$$y^2 = x^3 + k.$$

This work represents a remarkable discovery, long sought for in vain.

The work is naturally too difficult and intricate to be discussed here, but it may be of interest to mention that the approach to the Diophantine equation is different from that based on the Thue–Siegel–Roth theorem, outlined earlier. Instead of the Diophantine approximation properties of one algebraic number, one has to use the Diophantine approximation properties of the logarithms of several algebraic numbers.

NOTES

A good introduction to Diophantine equations is L. J. Mordell, *Diophantine Equations* (Academic Press, London, 1969). For more about Diophantine equations, see Nagell, or the more advanced monographs by Th. Skolem, *Diophantische Gleichungen* (Springer, 1937; reprinted by Chelsea Publ. Co., New York, 1950) and by Z. I. Borevich and I. R. Shafarevich, *Number Theory* (Academic Press, London, 1966).

The most remarkable general result hitherto proved is one that is due to Siegel; this gives a necessary and sufficient condition for an equation of the form $f(x,y) = 0$, where f is an irreducible polynomial, to have infinitely many solutions in integers x,y. See Skolem, ch. 6, §8.

§3. For the equation $ax^2 + by^2 = cz^2$, see also L. J. Mordell, *Monatshefte für Math.*, 55 (1951), 323–7.

There is a theorem of Dickson which states that if the equation $ax^2 + by^2 = cz^2$ is soluble, where a, b, c are square free and relatively prime in pairs, then every integer is representable in the form $ax^2 + by^2 - cz^2$. Thus from the example in the text it follows that every integer is representable in the form $41x^2 + 31y^2 - z^2$.

For an interesting account of the various methods which have been devised for equations of the form $y^2 = x^3 + k$, see L. J. Mordell, *A Chapter in the Theory of Numbers* (Cambridge, 1947).

§4. For an account of Fermat's last theorem see L. J. Mordell, *Three Lectures on Fermat's Last Theorem* (Cambridge, 1921), H. M. Edwards, *Fermat's Last Theorem: A Genetic Approach to Algebraic Number Theory* (Springer, 1977) and P. Ribenboim, *13 Lectures on Fermat's Last Theorem* (Springer, 1979). For numerical evidence see S. S. Wagstaff Jr, *Math. Comput.*, 32 (1978), 583–91. See also Guy, section D.1.

§5. See Dickson's *History*, vol. II, ch. 21, and K. Mahler, *J. London Math. Soc.*, 11 (1936), 136–8. For the anecdote about Ramanujan, see Hardy's memoir in *Collected Papers of S. Ramanujan* (Cambridge, 1927), or *Proc. London Math. Soc.* (2), 19 (1921), xl–lviii. For the four cube problem, see H. W. Richmond, *Messenger of Math.*, 51 (1922), 177–86, and L. J. Mordell, *J. London Math. Soc.*, 11 (1936), 208–218. For recent progress, see the review in *Math. Reviews*, 34 (1967), 445 of a paper by V. A. Demjanenko.

It may be that the equation $x^3 + y^3 + z^3 = 3$ has only a finite number of integral solutions. The only known solutions are 1, 1, 1 and 4, 4, -5.

§6. Roth's theorem was published in *Mathematika*, 2 (1955), 1–20 and 168. Other versions and generalizations will be found in J. W. S. Cassels's *Introduction to Diophantine Approximation* (Cambridge Tracts, no. 45, 1957; reprinted by Hafner Press, New York), in LeVeque, vol. 2 and in K. Mahler's *Lectures on Diophantine Approximations* (Univ. of Notre Dame, 1961). The appropriate generalization to the simultaneous approximation of several algebraic numbers was found by W. M. Schmidt, *Acta Mathematica*, 125 (1970), 189–201. For a systematic development of Roth's and Schmidt's theorems see Schmidt's *Diophantine Approximation* (Springer, Lecture Notes in Math., no. 785, 1980). Various applications of these results to Diophantine equations will be found in Schmidt's book.

Baker's fundamental work appeared in *Phil. Trans. Roy. Soc.* A 263 (1968), 173–91 and 193–208. The basic ideas underlying Baker's work are given in their simplest form in his paper in *Mathematika*, 13 (1966), 204–16. For an example of the use of Baker's method in solving Diophantine equations see A. Baker and H. Davenport, *Quart. J. of Math.*,

(2) 20 (1969), 129–37. Baker's results have been extended and applied in many ways. For a systematic treatment see Baker, *Transcendental Number Theory* (Cambridge, 1975) and M. Waldschmidt, *Nombres Transcendants* (Springer, 1974).

Another useful tool in obtaining solutions of the equation $f(x, y) = 0$ is Runge's theorem, see *Quart. J. of Math.*, (2) 12 (1961), 304–12 (310).

VIII

COMPUTERS AND THE THEORY OF NUMBERS

In this chapter, we shall assume some basic familiarity with computing, but not with any particular language or machine. We have included brief arguments describing the running time of the various algorithms — the reader not familiar with the complexity theory of algorithms can skip these.

1. Introduction

The rapid development of electronic computers has meant that number-theoretic calculations, which were until recently impossible or extremely difficult, can now be performed routinely on quite modest computers, even on home computers or programmable calculators. Gauss' childhood feat of computing $1+2+\cdots+100$ in his head can now be done in fractions of a millisecond. The comparison is not completely straightforward, as it is believed that Gauss actually achieved this feat by inventing the formula for the sum of the first n numbers, as $n(n + 1)/2$, and just substituted $n = 100$ in this — a feat which computers find more difficult, though far from impossible.

Computer designers typically provide computers capable of manipulating whole numbers up to a certain limit, often $32767 = 2^{15} - 1$ or $2147483647 = 2^{31} - 1$. For major computations, such as the recently-computed factorizations

$$2^{484} + 1 = 4994797680505587570210555567669066608919775702826395384137465113540059478211162499219248976490158715385572308979425059663271676108686125649006428 17$$

$$= 17 \times 353 \times 209089 \times 33186913 \times 1251287137 \times 2931542417 \times 38608979869428210686559330362638245355335498797441 \times 84694409197705740057696939084347325062258739942360 85602665729,$$

$10^{142} + 1 = 101 \times 569 \times 7669 \times 380623849488714809\times$

$\quad 77169265188335087786895085504941\times$

$\quad 93611382287513950329431625811490669\times$

$\quad 8251988265906196670876248348671944663928843044608 1$

or

$2^{463} + 1 = 238170513177184465895202425368741325817001 20107002$

$\quad 03819930387084675118819289982315155262834978860451 6$

$\quad 29506630799413011852606182616644504780 9$

$\quad = 3 \times 2356759188941953 \times 76834966209858049526107\times$

$\quad 88119307925269041107418404833666787\times$

$\quad 4975326045514038000659718805165333685595913106186792$

$\quad 454026995210139,$

where the final computations involved the factorization of a 111-digit product of two primes, a 116-digit product of three primes and a 101-digit product of two primes respectively, it is quite clear that the ideas set out in chapter I, and the use of the computer manufacturers' limited range of integers, will not suffice.

Just as we can handle numbers greater than 9 in the decimal system by the use of multi-digit numbers (such as 12 or 561) and of techniques such as "long multiplication" and "long division", we can do the same on a computer, and, if the maximum number provided by the computer manufacturer is 2147483647, we can divide our large integers up into "digits" base 10000, say, and handle these in ways similar to long multiplication and long division. We need to use a base B such that $(B-1)^2$ is representable on our computer, since the product of two "digits" can be as large as $(B-1)^2$. This tends to make the "digits" be smaller than we would like, and hence the numbers have more "digits" than might seem possible. Fortunately, many computer manufacturers actually provide instructions which multiply two numbers and produce a double-length result, and instructions which divide double-length numbers by single-length numbers, but, unfortunately, high-level computer languages tend not to provide access to these facilities, and it is often necessary to resort to machine code programming. Whilst substantial ingenuity is required to get the details right and the programs as fast as possible, the methods are fundamentally as we have outlined them.

There are other methods, and a flourishing branch of computer science explores questions such as "what is the fastest way of multiplying two large integers". However, the definition of "large" in that context would probably not stoop to include the numbers we have written above, which computer scientists would regard as "medium-sized", if not "small".

The development of computers has done more than provide tools for number theory. It has also provided applications for number theory, to the point where a working knowledge of elementary number theory is considered essential for a computer scientist. There are many of these applications. As a trivial one, we mention that $355/113$ is a perfect floating-point approximation to π on most 32-bit computers because it is obtained by truncating the continued fraction

$$\pi = 3 + \frac{1}{7+}\frac{1}{15+}\frac{1}{1+}\frac{1}{292+}\cdots$$

before the 292 term, so that the error is less than $1/(115^2 \times 292) = 1/3861700$ — see IV.5 and IV.7. A less trivial application is to be found in the design of so-called random number generators, which is outlined in section 3. Congruences are fundamental to the design of hash tables, which are one of the most efficient ways of storing information for rapid retrieval.

But the most important applications of number theory to computing are in the area of public-key cryptography, which enables two people to share a secret, or one of them to verify that the other person is really who he claims to be, *without* pre-arranged codebooks. As the use of computers spreads further, from the banks to electronic transfer at the supermarket or shop, techniques such as this will be needed to combat the possibilities of fraud. These techniques are outlined in the last two sections of this chapter.

2. Testing for primality.

Many of the subjects that we shall discuss later require the use of large primes, often large "random" primes: random in the sense that they have no particular structure, and are not easy to guess, or to find in standard tables of large primes. For example, it is comparatively easy to tell if a large Mersenne number (one of the form $2^n - 1$) is prime or not: n has to be prime, and then there are the Lucas-Lehmer tests which will prove whether or not $2^n - 1$ is prime. Most of the very large primes which are known today are of this form. Regrettably, the special properties which make it easy to show that they are prime also make it easy to attack many of the codes based on large primes: mathematics rarely gives us something for nothing.

How can we tell if a large random number is prime? Fermat's theorem (II.3), that

$$x^{p-1} \equiv 1 \pmod{p}$$

for all integers x not congruent to 0, can often show that a number is not prime. For example, we can show in this way that 10 is not prime, by observing that

$$3^9 \equiv 3^4 \times 3^4 \times 3 \equiv 81 \times 81 \times 3 \equiv 3 \pmod{10},$$

and hence the pair $x = 3, p = 10$ would be a counter-example to Fermat's theorem if 10 were prime. Since the theorem is true, 10 cannot be prime.

This method can be used easily, and takes only a small amount of computer time, to show that numbers with hundreds of digits are not primes. To do this, we need to be able to compute x^{p-1} (mod p) rapidly. A preliminary remark is in order here: we must not first compute the integer x^{p-1}, and then reduce it to the modulus p, for this number would be totally outside the range of computability; rather we must work to the modulus p throughout the computation of x^{p-1}. For the computation of x^{p-1} (mod p), or more generally any x^k (mod p), we observe that, if k is even, then $x^k = (x^2)^{k/2}$, while if k is odd, say $k = 2l + 1$, then $x^k = x(x^2)^l$. At the expense of one or two multiplications, we have reduced the problem of computing x^k (mod p) to a similar problem with a value of k which is half of what it was. Hence the number of multiplications required by this method of *repeated squaring* is somewhere between $\log_2 k$ and $2\log_2 k$.

However, can we use this method to show that a number is prime? We recall the definition of $\phi(n)$ from chapter II — it is the number of numbers less than or equal to n and relatively prime to n. Euler's theorem (II.3) states that $x^{\phi(n)} \equiv 1$ (mod n) if x is relatively prime to n. In II.4, we showed that ϕ is a multiplicative function, and that

$$\phi(q_1^{a_1} q_2^{a_2} \ldots) = \phi(q_1^{a_1})\phi(q_2^{a_2})\ldots = q_1^{a_1-1}(q_1 - 1)q_2^{a_2-1}(q_2 - 1)\ldots.$$

Define $\hat{\phi}(q_1^{a_1} q_2^{a_2} \ldots)$ to be the least common multiple, rather than the product, of $\phi(q_1^{a_1}) = q_1^{a_1-1}(q_1 - 1)$, $\phi(q_2^{a_2}) = q_2^{a_2-1}(q_2 - 1), \ldots$. Then, for each of the factors $q_i^{a_i}$ of n and for x relatively prime to n, we deduce that $x^{\phi(q_i^{a_i})} \equiv 1$ (mod $q_i^{a_i}$), and so $x^{\hat{\phi}(n)} \equiv 1$ (mod $q_i^{a_i}$). It then follows that $x^{\hat{\phi}(n)} \equiv 1$ (mod n).

If we were unlucky enough to have a non-prime number n such that $\hat{\phi}(n)$ divides $n-1$, then every x relatively prime to n would have the property that $x^{n-1} \equiv 1$ (mod n), and, unless we were lucky enough to choose an x which had a factor in common with n, we would not be able to use the Fermat test to detect that n is not prime. Such numbers, though rare, actually do exist — they are called *pseudo-primes* or *Carmichael numbers*. The smallest such is $561 = 3 \times 11 \times 17$. So $\hat{\phi}(561) = lcm(3 - 1, 11 - 1, 17 - 1) = lcm(2, 10, 16) = 80$, which does divide 560. To illustrate the problem that these numbers can cause, let us try to show that 561 is not prime by looking at 2^{560} (mod 561). We get the following table of powers of two to the modulus 561, using the method of repeated squaring outlined above:

$$2^{35} \equiv 263$$
$$2^{70} \equiv 166$$
$$2^{140} \equiv 67$$
$$2^{280} \equiv 1$$
$$2^{560} \equiv 1.$$

However, although Fermat's theorem does not prove that 561 is not prime, we can prove that it is not prime by using Lagrange's theorem, that a polynomial of degree n has at most n solutions to a prime modulus (II.7). Consider the polynomial $x^2 - 1$. This certainly has solutions $x \equiv 1$ and $x \equiv -1$, but, to the modulus 561, it also has the solution $x \equiv 2^{140} \equiv 67$ from the table above. Since it is a polynomial of degree two with three solutions, Lagrange's theorem would be contradicted if 561 were a prime, and so we can conclude that 561 is definitely not a prime. In fact, we can also determine a partial factorization: $\gcd(67 - 1, 561) = 33 = 3 \times 11$, whilst $\gcd(67 + 1, 561) = 17$. This technique works because, to any modulus which is a prime factor of 561, 67 is a square root of 1, so must be congruent to 1 or -1 to that prime modulus.

Rabin made use of this idea, that we often see a contradiction either of Fermat's theorem or of Lagrange's theorem applied to the polynomial $x^2 - 1$, to produce a procedure which, when given a prime, will always say "probably prime", and, when given a non-prime, will say "probably prime" with probability at most $\frac{1}{4}$, the rest of the time it will prove that the number is composite. Let us now explore Rabin's method, assuming that n is a number whose primality we wish to investigate. If n is prime, then $x^{n-1} \equiv 1$ (mod n) for all non-zero x. Choose such a non-zero x (in practice one also avoids $x \equiv \pm 1$): this choice provides the random element implicit in the statements about probability made at the beginning of this paragraph. We intend to compute x^{n-1} (mod n) by repeated squaring, but we have to do this in a particular order. Write $n - 1$ as $2^l m$, where m is odd, and compute x^{n-1} as $(x^m)^{2^l}$, i.e. first compute x^m, then square it l times, thus computing $x^{2m}, x^{4m}, \ldots, x^{2^l m}$, all to the modulus n.

(a) If $x^m \equiv 1$ (mod n), then we terminate, saying "n is probably prime", since neither Fermat's theorem nor Lagrange's theorem is violated.

(b) If any of $x^m, x^{2m}, x^{4m}, \ldots, x^{2^l m} \equiv -1$, then again we terminate, saying "n is probably prime", for the same reason as before.

(c) If any of $x^{2m}, x^{4m}, \ldots, x^{2^l m} \equiv 1$, say $x^{2^k m} \equiv 1$, then we terminate, saying "n is definitely not prime". We now have a counter-example to

Lagrange's theorem, since $x^{2^{k-1}m}$ is a square root of unity, and it is not 1 (otherwise we would have detected this in clause (a), or in this clause for a smaller value of k) or -1, which would be detected by clause (b). In this case, as in the example of 561 earlier, we can factorize n by looking at $\gcd(x^{2^{k-1}m} \pm 1, n)$.

(d) If we get to the computation of $x^{2^{l}m}$ without terminating, we can say that "*n* is definitely not prime", since $x^{2^{l}m} \equiv 1$ would have been detected in previous steps, and $x^{2^{l}m} \not\equiv 1$ contradicts Fermat's theorem. However, we have no information about the potential factors of n.

In practice, this algorithm can be run on numbers of a thousand decimal digits quite quickly. It can be argued, though, that whilst the answer "*n* is definitely not prime" is certainly correct (even though no factor of n has been exhibited), the answer "*n* is probably prime" is not certain enough. Perhaps we could get "*n* is probably prime" 10 times for 10 different choices of x, even for a non-prime number.

The reply to this argument is provided by the following theorem of Rabin: *for any non-prime n, at most 25% of the possible values of x will reply "n is probably prime"*. For $n = 9$, the x-values 1 and -1 both say "9 is probably prime", but none of the six other possible values (remembering that $x \equiv 0$ is excluded) does, so that it is possible for 25% of the x to give the wrong response. This means that, if we try ten different random values of x, and get the reply "*n* is probably prime" for all of them, then either n really is prime, or we have observed a one-in-a-million (more accurately, one-in 1,048,576) freak event of getting an unlucky number every time. If even this level of certainty does not suffice, then we note that 20 different values of x will give us a one-in-a-billion chance of being wrong, and so on. It should be noted that, for the vast majority of composite numbers n, very few of the possible values of x will reply "*n* is probably prime".

Whilst we do not intend to analyse the running times of these algorithms in detail, we note that a single application of Rabin's algorithm will require at most $2 \log_2 n$ multiplications, all of numbers less than n, which are to be carried out to the modulus n. The time taken to perform such a calculation, by ordinary "long multiplication" methods, is proportional to the square of the number of digits, since every digit of the multiplier is multiplied by every digit of the multiplicand. Since the number of (binary) digits is $\log_2 n$, the total cost is proportional to $\log_2^3 n$. While faster methods of multiplication are known, they are not generally used for numbers of the size common in cryptographic uses of prime numbers.

How would we actually *prove* that a number n is prime? The simplest way is to exhibit a number x such that $x^{n-1} \equiv 1 \pmod{n}$, but that

$x^{(n-1)/d} \not\equiv 1$ (mod n) for all prime divisors d of $n - 1$, in other words a primitive root to the modulus n (III.1). This would imply that all the numbers $x, x^2, \ldots, x^{n-1}$ are distinct to the modulus n, and since they are all relatively prime to n, it follows that every number between 1 and $n - 1$ is relatively prime to n, i.e. that n has no proper factors. Such a number x, together with a factorization of $n - 1$, could be regarded as a *certificate* that n is prime, since the associated proof can easily be checked. Of course the factorization of $n - 1$ would have to be accompanied by certificates that all the factors there are primes, and so on. The difficulty of producing such a certificate is not, generally speaking, the labour of finding x, for there are many such x (in fact, $\phi(n-1)$ of them — see III.1), but rather the difficulty of factoring $n-1$. There has recently been much research into alternative forms of certificate that will be easier to find or to check, but in general the mathematics involved in these alternative certificates is rather deep. The largest number for which such a certificate has been produced by general-purpose methods has 1065 digits.

3. *"Random" Number Generators*

There are many uses in computing for "random" numbers of some kind. We have seen one in the previous section, where we wished to take various values of x "at random" to see whether n is or is not prime. Many kinds of computer simulation rely on random numbers, just as games rely on the toss of a coin or the roll of dice. For some applications, such as the determination of prizes in Premium Bonds or lotteries, it is necessary for the numbers to be truly unpredictable, and resort must be had to some unpredictable physical process, rather than to arithmetic.

For many purposes, however, complete unpredictability is not so important, provided that the sequence of random numbers is "not too regular". What is more important is computational efficiency. This leads to the study of so-called *pseudo-random* numbers, where each number actually depends on the previous one, but in a manner that does not destroy the useful properties of the sequence. It is common to regard such a sequence as consisting of numbers to the modulus n, just as the numbers on a die can be viewed as being to the modulus 6. In practice, n is often chosen to be related to the properties, especially word-size, of the actual computer being used. Surely it should be easy to design a method which, given some number x_1 to the modulus n, scrambled it to produce x_2, then scrambled that to produce x_3, and so on.

One of the first such methods suggested was the *mid-square* method. This relies on squaring the numbers, and then taking the middle half of the square as the next number. If n were 10000 (probably too small in practice, but large enough to illustrate the point), so that the "middle half" of the square of a number is obtained by the deleting the first two and last two of the eight digits, and x_1 were 4321, we would see

$$x_1^2 = 4321^2 = 18671041, \text{ so } x_2 = 6710;$$
$$x_2^2 = 6710^2 = 45024100, \text{ so } x_3 = 241;$$
$$x_3^2 = 241^2 = 58081, \text{ so } x_4 = 580;$$
$$x_4^2 = 580^2 = 336400, \text{ so } x_5 = 3364;$$
$$x_5^2 = 3364^2 = 11316496, \text{ so } x_6 = 3164;$$
$$x_6^2 = 3164^2 = 10010896, \text{ so } x_7 = 108;$$
$$x_7^2 = 108^2 = 11664, \text{ so } x_8 = 116;$$
$$x_8^2 = 116^2 = 13456, \text{ so } x_9 = 134;$$
$$x_9^2 = 134^2 = 17956, \text{ so } x_{10} = 179;$$
$$x_{10}^2 = 179^2 = 32041, \text{ so } x_{11} = 320;$$
$$x_{11}^2 = 320^2 = 102400, \text{ so } x_{12} = 1024;$$
$$x_{12}^2 = 1024^2 = 1048576, \text{ so } x_{13} = 485.$$

There is clearly a strong tendency for one small number to be followed by another. It is also possible for the system to get stuck at 0, or at the short loop 6100, 2100, 4100, 8100, 6100, ..., as indeed this system does, with $x_{68} = 6100$. In fact, this is not so surprising, since methods chosen "at random" turn out not to be random enough, as the next example illustrates.

There is a well-known "paradox" (actually an illustration that the laws of probability do not behave as we naïvely expect) that, if we have 23 or more people together in a room, it is more likely than not that two of them have the same birthday. The proof of this is easy if we ignore the existence of leap years, as we shall do, and a little more complex if we take them into account. If no two of the people have the same birthday, then the first person to enter the room could have been born on any day (probability 365/365), the second can have any birthday except the first person's (probability 364/365), the third can have any birthday except either of those of the first two people (probability 363/365), and so on, which gives us a cumulative probability for 23 people in the room of

$$\frac{365}{365} \times \frac{364}{365} \times \frac{363}{365} \times \dots \times \frac{365 - 22}{365}$$

which works out to be

'369979785662179593401824991341667570443383351847256064
7509188326851535012542620742522314756326980590820312 5'

the numeric value of which is about 0.4927. Hence the probability that two do have the same birthday is about 0.5073, greater than one-half. The same general phenomenon occurs whatever the number of days in a year (or of other objects from which we are selecting). In fact, probability theory tells us that, if we are selecting from N possible objects, we expect a repetition after about $\sqrt{\pi N/2}$ selections, which for $N = 365$ gives 23·94: an excellent agreement with the calculation above. For $N = 10000$, as in the example of mid-square random number generation, we would expect a repetition within 125 elements, so finding it at $x_{72} = x_{68} = 6100$ is not too surprising.

Hence we need to think about our choice of method, rather than just choose one at random. What requirements do we wish our random sequence to have?

- We want a long period between repetitions. Ideally, if our sequence is of the form $x_{i+1} = f(x_i)$ (mod n), we would want x_i to take all possible values to the modulus n before repeating.
- We want our sequence to "look random". The repeated occurrence of small numbers in the mid-square method certainly does not look random. The following sequence: $x_{i+1} = 1 + x_i$ (mod n) satisfies the criterion of trying every value, but few people would claim that this is random.

The first criterion is amenable to, indeed it requires, arithmetic methods for its satisfaction, whereas the second one needs statistical methods for its precise formulation, and certainly for its satisfaction. We shall concentrate on the first, but the reader must bear in mind that whilst satisfying the first criterion is definitely *necessary* to produce a good random number generator, it certainly is not *sufficient*. At the end of this section we shall give a few possible methods which are widely believed to satisfy both criteria.

One of the most popular methods of generating such pseudo-random numbers is the so-called *linear congruential method*:

$$x_{i+1} = (ax_i + c) \quad (\text{mod } n) \tag{1}$$

where x_{i+1} satisfies a linear congruence (in the sense of chapter II) in terms of x_i. We shall always use a and c in this sense for the rest of this section, and often use b to stand for $a - 1$. If we substitute equation (1) into the analogous equation giving x_{i+2} in terms of x_{i+1} we get

$$x_{i+2} = (ax_{i+1} + c) \quad (\text{mod } n)$$
$$= (a(ax_i + c) + c) = a^2 x_i + (a + 1)c \quad (\text{mod } n).$$

This process can clearly be continued, expressing x_{i+3} in terms of x_i and so on. If we use the algebraic identity

$$a^{j-1} + a^{j-2} + \cdots + a + 1 = \frac{(a^j - 1)}{a - 1} = \frac{(a^j - 1)}{b}$$

we get the concise expression

$$x_{i+j} = (a^j x_i + (a^j - 1)c/b) \pmod{n}. \tag{2}$$

This has the same form as (1), with a replaced by a^j and c replaced by $(a^j - 1)c/b$. Hence the view held by some programmers, that they can make a sequence which is "twice as random" by taking every alternate element of the sequence, is fallacious: the same sequence can be obtained by choosing different values of a and c. As we shall see later, it is generally not helpful to perform this transformation.

Let us now study the fundamental *arithmetic* question of choosing good linear congruential random number generators:

what values of x_1, a, c and n give the maximum period of the generator, i.e. cause every value to the modulus n to be taken before the sequence repeats?

It turns out that x_1 is not particularly important in this. Consider a similar sequence, but starting from 0 and with $c = 1$:

$$y_1 = 0 \qquad \text{and} \qquad y_{i+1} \equiv (ay_i + 1) \pmod{n}. \tag{3}$$

Then, as in (2) above, $y_k \equiv (a^{k-1} - 1)/b \pmod{n}$, whereas

$$\begin{aligned}
x_k &\equiv a^{k-1}x_1 + c(a^{k-1} - 1)/b \pmod{n} \\
&\equiv (by_k + 1)x_1 + cy_k \pmod{n} \\
&\equiv (x_1 b + c)y_k + x_1 \pmod{n}.
\end{aligned}$$

So if $x_1 b + c$ is relatively prime to n, the sequence of x_i has precisely the same period as that of the y_i. If $x_1 b + c$ is not relatively prime to n, then the sequence of x_i will have a shorter period: the same as that of the y_i taken to the modulus $n/\gcd(n, x_1 b + c)$.

We now need a technical result, which can be viewed as a generalization of Fermat's theorem (II.3). Let p be a prime and e be a natural number such that $p^e > 2$ (i.e. we are ruling out just one case: $p = 2$ and $e = 1$). Suppose that

$$x \equiv 1 \pmod{p^e} \qquad \text{and} \qquad x \not\equiv 1 \pmod{p^{e+1}}. \tag{4}$$

Then

$$x^p \equiv 1 \pmod{p^{e+1}} \qquad \text{and} \qquad x^p \not\equiv 1 \pmod{p^{e+2}}. \tag{5}$$

We note that the case $p^e = 2$ and $x = 3$ shows that $p^e = 2$ has to be excluded. The proof is similar to Leibniz' proof (II.3) of Fermat's theorem.

We can write $x \equiv 1 + qp^e \pmod{p^{e+1}}$ where $q \not\equiv 0 \pmod{p}$. Now expand $(1 + qp^e)^p$ by the binomial theorem, to obtain

$$1^p + p1^{p-1}qp^e + \frac{p(p-1)}{2}1^{p-2}(qp^e)^2 + \underbrace{\frac{p(p-1)(p-2)}{6}1^{p-3}(qp^e)^3 + \cdots}_{\text{divisible by } p^{e+2}}.$$

Unless $p = 2$, we see that $\frac{p(p-1)}{2}1^{p-2}(qp^e)^2$ is divisible by p^{e+2}: a contribution of p from the binomial factor $\frac{p(p-1)}{2}$ and a contribution of at least p^{e+1} from the p^{2e}. If $p = 2$, we know that $e > 1$, so the p^{2e} term contributes at least p^{e+2}. In either case, therefore, all terms are divisible by p^{e+2} except for the first two. Hence $x^p \equiv 1 + qp^{e+1} \pmod{p^{e+2}}$, which proves (5).

Now let us consider the special case of a generator with $n = p^e$. The case $n = 2$ is trivial, for the sequence of maximal length is $0, 1, 0, 1, \ldots$. This illustrates the folly of thinking that "random coin tossing" can be obtained by calculation to the modulus two: we should use a much larger modulus n, the largest we can, for the random number generator, and later reduce the answers to the modulus two. However, it is not a good idea to compute the answers to the modulus two by taking the remainder of the sequence (mod n) on division by two, since, if n is odd, we shall have a slight bias in favour of 0, whilst if n is even, we shall effectively have a sequence to the modulus two, and the period will be at most two. The correct solution for even n is to divide the sequence (mod n) by $n/2$, and consider the quotient. With luck, though this has to be checked, the sequence thus obtained will have period the same as the original sequence. For odd n, we divide by $(n - 1)/2$ and take the quotient if it is 0 or 1, if it is two we take the next member of the sequence and divide it by $(n - 1)/2$.

We shall prove that the sequence has maximal period length if, and only if, the following three conditions are satisfied:

(1) p divides b;

(2) if $p = 2$, then 4 divides b;

(3) p does not divide c.

If the x_i are to have maximal period length, then the y_i must have maximal period length. Since $y_{k+1} = (a^k - 1)/b$, we must prove that this first attains the value 0 (to the modulus n) when $k = n$. If $a \not\equiv 1 \pmod{p}$, then $a^{\phi(n)} \equiv 1 \pmod{n}$, and so $(a^{\phi(n)} - 1)/b \equiv 0 \pmod{n}$; thus the sequence attains 0 too soon. This argument will not work when $a \equiv 1 \pmod{p}$, for then we cannot simply divide by b, since $b \equiv 0 \pmod{p}$. So we have proved that condition (1) must hold. If condition (2) does not hold, then p is two and $a \equiv 3 \pmod{4}$. But then $a^2 \equiv 1 \pmod{8}$ and, by a repeated application of (4) and (5) above, $a^{2^2} \equiv 1 \pmod{2^4}$ and so on;

thus $a^{2^{e-1}} - 1 \equiv 0 \pmod{2^{e+1}}$. Since two divides $a - 1$ but four does not, we can divide this congruence by $a - 1$ at the cost of writing it to the modulus 2^e, and obtain $(a^{2^{e-1}} - 1)/b \equiv 0 \pmod{2^e}$, which shows that the sequence repeats at 2^{e-1} rather than at 2^e. We have shown that conditions (1) and (2) are necessary if the sequence of y_i is to have maximal length.

We now have to show that, if (1) and (2) are satisfied, then the sequence of y_i does actually have maximal length. If $a \equiv 1 \pmod{p^e}$ the sequence certainly does have maximal length, since it is the sequence 0,1,2,3,.... So suppose that $a \equiv 1 \pmod{p^f}$, but that $a \not\equiv 1 \pmod{p^{f+1}}$, for some value of f less than e. Then by repeated application of (4) and (5), we see that $a^{p^e} \equiv 1 \pmod{p^{f+e}}$, but $a^{p^e} \not\equiv 1 \pmod{p^{f+e+1}}$. Hence the sequence repeats (not necessarily for the first time!) after p^e steps, since $a^{p^e} - 1 \equiv 0 \pmod{p^{f+e}}$, and dividing this congruence by $a - 1$, which is divisible by p^f, means writing it to the modulus p^e rather than p^{f+e}. Hence the actual period length must be a factor of p^e, since otherwise the remainder on dividing p^e by the actual period length would also be a period length. Therefore actual period length is p^g for some g. This g has to be equal to e, since for all smaller values of g, we do not have p^{f+e} dividing $a^{p^g} - 1$.

Thus conditions (1) and (2) are both necessary and sufficient for the y_i to have maximal period length. What about the x_i? We observed just after equation (3) that if, and only if, $x_1 b + c$ is relatively prime to n, the sequence of x_i has precisely the same period as that of the y_i. Since $n = p^e$ and p divides b, this condition is the same as requiring p not to divide c, i.e. condition (3).

We must now consider the case of general n, rather than the special case $n = p^e$. We shall show that the sequence has maximal period length if, and only if, the following three conditions are satisfied:

(1′) p divides b, for all p dividing n;
(2′) if two divides n, then four divides b;
(3′) n and c are relatively prime.

If the sequence is to have maximal period length to the modulus n, then, by the Chinese Remainder Theorem (II.4), it must have maximal period length to the modulus p^e for each p^e dividing n, since the period to the modulus n will be the least common multiple of the periods to the moduli p^e. But (1′), (2′) and (3′) are equivalent to requiring (1), (2) and (3) for each such p^e.

In practice, some conditions slightly stronger than (1′), (2′) and (3′) are necessary to ensure that the sequence does not have bad statistical properties. If a random number generator is going to be used extensively, then proper statistical tests should be performed on the sequences generated.

- The modulus n should be as large as practicable: generally the computer's word size is the most suitable choice.

- In addition to (1'), (2') and (3'), if two divides n, then we should choose $a \equiv 5 \pmod 8$, and if 10 divides n, then we should choose $a \equiv 21 \pmod{200}$.

- a should be chosen between $n/10$ and $9n/10$ and, subject to the previous congruence conditions, should not have a simple pattern of binary or decimal digits. For the common case of modulus $4294967296 = 2^{32}$, a set of parameters which have good statistical as well as arithmetic properties is $a = 2147001325$, $c = 715136305$.

4. Pollard's Factoring Method

Pollard used the observation of the last section, that "random" methods are not random enough, to produce an ingenious factoring algorithm, where the average running time for factoring n is proportional to $n^{1/4} \log^2 n$, whereas the algorithms sketched in I.9 take, in general, time proportional to $n^{1/2}$ or worse. It is worth noting that this method should only be applied to numbers which are known not to be prime — fortunately Rabin's algorithm of section 2 supplies us with an efficient method for deciding this.

Let us suppose that we have some procedure f to the modulus n, which, given a number x_i, returns another number $x_{i+1} = f(x_i)$. A method which works well in practice is to take $x_{i+1} = x_i^2 + 1 \pmod n$. If this method is "sufficiently random", then the probability theory quoted in the previous section says that it will repeat, on average, after $\sqrt{(\pi n/2)}$ different values of i. In fact, the particular formula mentioned above will repeat somewhat sooner: since x_i^2 has to be a quadratic residue (III.3) to the modulus n, not all values to the modulus n will be used. If n were prime, only $(n+1)/2$ different values of $x_i^2 + 1 \pmod n$ are possible (corresponding to the $(n-1)/2$ proper quadratic residues and the special case of $x_i = 0$).

If p is a factor of n, we then expect a repetition to the modulus p after about $\sqrt{(\pi p/4)}$ selections. However, the first difficulty is that p is unknown: the aim of factoring n is to discover p. This problem can be circumvented by observing that a repetition to the modulus p, say $x_i \equiv x_j \pmod p$, means that $\gcd(n, x_i - x_j)$ will be non-trivial. The second difficulty is that comparison of each x_i with each x_j (where comparison means the computation of $\gcd(n, x_i - x_j)$) would take about $\pi p/32$ such computations, and this would probably not be faster than the trial division methods of I.9. We need some way to detect repetitions more rapidly.

This is provided by what is called Pollard's "rho" method, based on observing that a repeating sequence looks like the Greek letter rho, or ρ, in that there is an irregular part at the front of the sequence, corresponding to the tail of the ρ, followed by a circle which repeats indefinitely. This follows from our definition of the x_i: if $x_i = x_j$, then $x_{i+1} = f(x_i) = f(x_j) = x_{j+1}$.

Pollard's method relies on comparing:

x_1 with x_2;

x_2 with x_3 and x_4;

x_4 with $x_5, \ldots, x_8$;

x_8 with $x_9, \ldots, x_{16}$

and so on. Suppose the first repetition to the modulus p occurs when x_i is equal to some earlier x_j. In terms of the "rho" picture, this means that $x_1, \ldots, x_{j-1}$ lie on the tail, x_j is where the tail joins the main body, and $x_{j+1}, \ldots, x_{i-1}$ lie round the circle. Suppose t is the first power of 2 larger than (or equal to) i. Then, as $x_j = x_i$, we have $x_{j+1} = x_{i+1}$ and so on, until we obtain $x_t = x_{t+i-j}$. Since t is at least as large as i, $t + i - j$ must lie between t and $2t$, and so our method of comparison ensures that we shall compare x_t with x_{t+i-j}. This comparison involves the computation of $\gcd(n, x_t - x_{t+i-j})$, which will be divisible by p because x_{t+i-j} is a repetition of x_t. The only thing that can go wrong is that x_{t+i-j} could conceivably also be a repetition of x_t for the other factors as well, i.e. it could actually be a repetition to the modulus n, and then the gcd would just be n, and we would have learnt nothing about the factorization of n. In practice this is extremely rare: should it happen, we can restart the method at a different value of x_1, or, preferably, with a different choice of f.

A couple of practical remarks are called for. The first is that the repetition may well be detected earlier: if the power of two before i, say t', is larger than both j and $i - j$, then the repetition will be discovered on comparing $x_{t'+i-j}$ with $x_{t'}$. Another practical point is that the key computations consist of the $\gcd(x_t - x_i, n)$. Since gcd is a comparatively expensive computation, it may make sense to aggregate a few of these computations, so that we compute, say, $\gcd((x_t - x_i)(x_t - x_{i+1}), n)$, then $\gcd((x_t - x_{i+2})(x_t - x_{i+3}), n)$, and so on, thus doing only half as many gcd computations. Of course, there is a slightly greater risk that the gcd will be n, but we could then try each gcd separately if this were to happen.

At the beginning of this section, we stated that the running time of Pollard's rho algorithm was proportional to $n^{1/4}$, but in fact we have proved something rather better: it is proportional to $p^{1/2} \log^2 n$, where p is the factor it finds, and the $\log^2 n$ term comes from the manipulation of numbers to the modulus n. This means that it is an excellent supplement to the "trial division" methods of chapter I for finding rather small, but not very small, factors of large numbers. In the case of the factorization of $2^{484} + 1$ given at the beginning of this chapter, the factors 17, 353, and possibly even 209089 could be found by trial division (say by all the primes up to a million); however the next three factors would be extremely expensive to find by trial division, but were found reasonably easily by Pollard's algorithm, since

100,000 iterations of Pollard's algorithm ought to find factors less than about 10^{10}. However, it would require something like 10^{27} (a thousand million million million million) iterations to find the remaining factors, so it is clearly not a solution to all our factoring problems.

5. Factoring large numbers

How should we factor a large number N? The first step is to look for small factors, typically by trying every divisor up to some bound such as 100,000. We could save some time by having a table of all the primes up to the bound, but this would take up space. A common compromise is to divide by two, three and then numbers congruent to 1 or 5 (mod 6). Once we have eliminated all the small factors, we can then see whether the number is prime: the method of section 2 is well-suited to this.

If the number is not prime, the method of section 2 will probably not have found any factors, and we shall be left in the tantalizing, but common, position of knowing that N is not prime, but not knowing its factors. We can then try some more advanced methods: for example 50,000 iterations of Pollard's method will probably find any factors less than 10,000,000,000. After each such factor is found, we have to test the remaining number for primality. If Pollard's method finds a factor larger than the square of the bound used for trial division, we should also test that this factor is actually prime, since there is a remote chance that it will not be.

There is a more advanced method than Pollard's rho method, which also has the advantage that its running time depends on the size of the factor found. To explain this dependence, we first need a definition: let $L(x)$ be a function such that $\log L(x) = \sqrt{\log x \log \log x}$, which means that $L(x)$ has the property that, as x increases, $L(x)$ increases more slowly than x, or $\sqrt{x}$, or $x^{1/3}$, or $x^{1/n}$ for any value of n. On the other hand, $L(x)$ increases more quickly than $\log x$, or $\log^2 x$, or $\log^n x$ for any value of n. It therefore provides an intermediate measure of growth: slower than any root of x, but faster than any power of $\log x$. This factoring method, which is based on the arithmetic of elliptic curves, a subject too technical for us to describe here, is expected to find a factor p of a number n in time proportional to $L^2(p) \log^2 n$, where the factor $\log^2 n$ comes from the fact that the algorithm is manipulating numbers to the modulus n, and the $L^2(p)$ term describes the inherent complexity of the algorithm. If n were the product of two similarly-sized numbers, then the running time of the elliptic curve algorithm would be proportional to $L(n) \log^2 n$. As with Pollard's algorithm, this method is well-suited to finding small factors of large numbers, but this method is suitable for finding larger factors than those which Pollard's algorithm can reasonably find. For example, a 30-digit number would require something like 10^{15} iterations of Pollard's algorithm, but more like 10^{11} iterations of the elliptic curve algorithm — a factor of ten thousand less. This method was used to find the factor 380623849488714809 of $10^{142} + 1$.

In practice, though, even the elliptic curve algorithm is not the most efficient one known. Following Fermat, we observed in I.9 that, if we know x and y such that $x^2 - N = y^2$, then $N = (x + y)(x - y)$. Searching for such x and y directly is only suitable if y is very small, i.e. if the two factors of N are very close together. Nevertheless, developments of this idea form the basis of the most advanced factoring algorithms known. First, we note that it is not necessary for N to be equal to $x^2 - y^2$: it is enough that $x^2 - y^2 \equiv 0$ (mod N) and that neither $x - y$ nor $x + y \equiv 0$ (mod N). So, we should look for non-trivial solutions to $x^2 \equiv y^2$ (mod N). Looking at random is unlikely to find such solutions: we need a way of constructing such solutions.

The basic method adopted is to find several numbers x_i such that x_i^2 is congruent to a relatively small number, to factorize these numbers, and to use these factorizations to find a combination of the x_i such that the square of their product, when reduced to the modulus N, is also a square. Consider as an example the number $N = 197209$. We can observe that $159316^2 \equiv 720 = 2^4 3^2 5$ (mod 197209) and that $133218^2 \equiv 405 = 3^4 5$ (mod 197209). Neither 720 nor 405, regarded as natural numbes, are squares, since each of them has an isolated factor of 5. But their product will be a square, since it is $2^4 3^6 5^2 = \left(2^2 3^3 5\right)^2 = 540^2$. So we have shown that $(159316 \times 133218)^2 \equiv 540^2$ (mod 197209), which reduces to $126308^2 \equiv 540^2$ (mod 197209). Since $\gcd(126308 - 540, 197209) = 199$ and $\gcd(126308 + 540, 197209) = 991$, we deduce the factorization $197209 = 199 \times 991$.

How could we have deduced that the numbers 159316 and 133218 had squares which were congruent to particularly small numbers? The continued fraction expansion of $\sqrt{197209}$ gives us a clue:

$$\sqrt{197209} = 444 + \frac{1}{12+} \frac{1}{6+} \frac{1}{23+} \frac{1}{1+} \frac{1}{5+} \frac{1}{3+} \frac{1}{1+} \frac{1}{26+} \frac{1}{6+} \frac{1}{2+} \frac{1}{36+} \cdots$$

Let q_n denote the n-th term in this continued fraction expansion, and let A_n/B_n denote the n-th convergent to $\sqrt{197209}$. By the theory of IV.6, the error $\left| \sqrt{197209} - \frac{A_n}{B_n} \right|$ is less than $1/B_n B_{n+1}$, which in turn is less than $1/q_{n+1}B_n^2$. So the convergents immediately preceding a large term are particularly good approximations, but all convergents are good approximations. If we write $A_n/B_n = \sqrt{197209} + e$, where we have shown that e is less than $1/B_n^2$, we can write $\left(A_n/B_n\right)^2 = 197209 + 2e\sqrt{197209} + e^2$, which means that $A_n^2 = 197209B_n^2 + 2e\sqrt{197209}B_n^2 + e^2 B_n^2$. If we write

$E = 2e\sqrt{197209}B_n^2 + e^2 B_n^2$, the previous equation becomes the congruence $A_n^2 \equiv E \pmod{197209}$, and E has to be less than $2\sqrt{197209}$. A good convergent is

$$444 + \cfrac{1}{12+} \cfrac{1}{6} = \frac{32418}{73},$$

when $E = 37$, small but unfortunately not a product of very small primes. The next convergent is

$$444 + \cfrac{1}{12+} \cfrac{1}{6+} \cfrac{1}{23} = \frac{750943}{1691},$$

and here the value of E is 720. So $750943^2 \equiv 720 \pmod{197209}$, which is the congruence $159316^2 \equiv 720 \pmod{197209}$ which we observed before. The convergent

$$444 + \cfrac{1}{12+} \cfrac{1}{6+} \cfrac{1}{23+} \cfrac{1}{1+} \cfrac{1}{5+} \cfrac{1}{3+} \cfrac{1}{1+} \cfrac{1}{26+} \cfrac{1}{6} = \frac{3143053051}{7077638}$$

gives rise to the congruence

$$3143053051^2 \equiv 3143053051^2 - 197209 \times 7077638^2 \equiv 405,$$

which reduces to $133218^2 \equiv 405 \pmod{197209}$.

There is nothing special about 197209, and the method can be applied to any integer known not to be prime. One possible drawback is that the continued fraction for $\sqrt{N}$ may repeat very rapidly: in this case we replace N by kN for some small k, and look at the continued fraction expansion of $\sqrt{kN}$. The choice of k can also affect the probability that a prime will divide E. Let us consider whether 5 divides E. Since $A_n^2 = kNB_n^2 + E$, we can write $A_n^2 \equiv kNB_n^2 + E \pmod 5$. We showed in IV.4 that A_n and B_n are always relatively prime, so there are 24 possible values for A_n and B_n modulo 5 — all combinations except $(0,0)$. If $kN \equiv 0 \pmod 5$, then only the four combinations with $A_n \equiv 0 \pmod 5$ will make $E \equiv 0 \pmod 5$, and then $E \equiv 0 \pmod{25}$ if, and only if, $kN \equiv 0 \pmod{25}$. If $kN \equiv \pm 1 \pmod 5$ (i.e. is a quadratic residue) then the eight combinations with $A_n^2 \equiv \pm B_n^2$ — two values for A_n for every non-zero value of B_n — will make $E \equiv 0 \pmod 5$. Conversely, if $kN \equiv \pm 2 \pmod 5$ (i.e. is a quadratic non-residue) then the multiplicative property of quadratic residues (III.3) means that $E \not\equiv 0 \pmod 5$.

Another important practical point is that we do not need to compute the convergents and then reduce the numerator and denominator modulo N: rather we can compute the numerator and denominator using the recurrence relations $A_m = q_m A_{m-1} + A_{m-2}$ and $B_m = q_m B_{m-1} + B_{m-2}$ (IV.4), but interpreting these to the modulus N, since we are only interested in the values of A_m and B_m to the modulus N. The production of the congruences $A_m^2 \equiv E$ (mod N) can be made sufficiently fast that almost all the time is consumed in factoring the E. The obvious strategy is to select a set of primes (generally the first n primes $p_1, \ldots, p_n$) and to see which E can be expressed as a product of powers of these primes and of the number -1, which we treat as if it were a prime for this process, and call p_0. A carefully-written trial division process is then used to perform the factorization.

Once we have sufficiently many congruences of the form

$$A_j^2 \equiv p_0^{e_{j0}} p_1^{e_{j1}} \cdots p_n^{e_{jn}} \quad (\text{mod } N),$$

we can start looking for a combination of the A_j such that the product of their squares is also congruent to a different square. This means that the exponent of every p_i in the product must be even. If we write $a_j = 1$ to indicate that A_j^2 will occur in the product, and $a_j = 0$ to indicate that A_j^2 will not occur in the product, then the exponent of p_i in the product is the sum $a_1 e_{1i} + \cdots + a_k e_{ki}$. The requirement that all these sums be even is equivalent to finding a non-trivial solution to a system of linear equations to the modulus two:

$$a_1 e_{10} + \cdots + a_k e_{k0} \equiv 0 \quad (\text{mod } 2)$$
$$a_1 e_{11} + \cdots + a_k e_{k1} \equiv 0 \quad (\text{mod } 2)$$
$$\cdots \qquad \cdots$$
$$a_1 e_{1n} + \cdots + a_k e_{kn} \equiv 0 \quad (\text{mod } 2).$$

There is one addition that can usefully be made to this scheme, the so-called "large prime variant". In this scheme, rather than insist that $A_j^2 \equiv p_0^{e_{j0}} p_1^{e_{j1}} \cdots p_n^{e_{jn}}$ (mod N), in other words that E has been factored completely, we allow one additional, larger, prime, so that $A_j^2 \equiv p_0^{e_{j0}} p_1^{e_{j1}} \cdots p_n^{e_{jn}} Q_j$ (mod N) is also permissible, with Q_j a large prime. The obvious definition of "large" in this context is "larger than p_n but smaller than $p_n(p_{n+1})$, since any number in this range left after trial division by $p_1, \ldots, p_n$ has to be prime. Potentially, this generates congruences faster than this simpler method of the previous paragraph, but the corresponding system of linear equations might appear to have many more terms in each equation, since we have almost squared the number of primes available. However, at most one "large" prime occurs in each equation — a specialist in linear equations would say that these equations are very sparse. We can make use of this sparsity in the following way: as the congruences are generated, they are stored according to the value of Q_j occurring in them, if any. If we discover two congruences with the same large prime in them, say $A_j^2 \equiv p_0^{e_{j0}} p_1^{e_{j1}} \cdots p_n^{e_{jn}} Q_j \pmod{N}$ and $A_i^2 \equiv p_0^{e_{i0}} p_1^{e_{i1}} \cdots p_n^{e_{in}} Q_i \pmod{N}$ with $Q_j = Q_i$, we can construct an equation without a large prime, viz.

$$\left(\frac{A_i A_j}{Q_i}\right)^2 \equiv p_0^{e_{i0}+e_{j0}} p_1^{e_{i1}+e_{j1}} \cdots p_n^{e_{in}+e_{jn}} \pmod{N}.$$

where the division is to be interpreted as taking place to the modulus N in the sense of II.2 — if this division were to fail, we would obtain a factor of N. When we have accumulated enough equations involving only the p_i, either obtained via the technique just outlined or directly because E factored completely, we solve the linear equations to the modulus two as before.

The time taken by this algorithm is rather hard to analyze, since it depends on the choice of k, and of n, the number of primes in the factor base, as well as on the details of the algorithm implemented. Too small a value for n will mean that very few congruences will give rise to equations, whilst too large a value for n will increase the time taken to factorize a given E, and the time required for the solution of the linear equations to the modulus two. In practice, the solution of the equations, and the computer memory required to store the equations, is often the limiting factor. If n is chosen such that $\log n = \frac{1}{2}\sqrt{\log N \log \log N}$, which seems to be the best value from the point of view of theoretical analysis, then it can be shown that the running time of the basic algorithm is at most proportional to $L(N)^2$. In practice, and with the large prime variant, it seems to be proportional to $L(N)$.

There is another way of generating these congruences, known as the *quadratic sieve* method, which does not rely so heavily on trial division: instead we construct congruences $A^2 \equiv B \pmod{N}$ where we know, not only that B is small, but that it has many small prime factors. We may assume that N, the number we wish to factor, has no small prime factors. Let M be a whole number as close as possible to $\sqrt{N}$, and let $Q(x)$ be the function $(M + x)^2 - N$. When x is a small integer, this is of size about $2x\sqrt{N}$, and therefore is relatively likely to factor into small integers. The ingenious part about the quadratic sieve is that we can state which primes will divide the various values of $Q(A)$. Two clearly divides the even ones, i.e. exactly half of them.

How many of them does 3 divide? If the quadratic residue symbol $(N|3)$ (III.3) is -1, then $N \equiv (M + x)^2 \pmod{3}$ is impossible. Conversely, if $(N|3) = 1$, then N has two square roots to the modulus 3, and 3 will divide every $(M + x)^2 - N$ such that $M + x \equiv \pm 1$ (mod 3), i.e. two-thirds of the possibilities rather than the one-third one might expect. The argument works for any prime p: if $(N|p) = -1$ then p divides no values of $(M + x)^2 - N$, whilst if $(N|p) = 1$, then N has two square roots to the modulus p, say $\pm a$, and p divides those values of $(M + x)^2 - N$ for which $M + x \equiv \pm a$. So the values of x for which $(M + x)^2 - N$ is divisible by p form two arithmetic progressions, and a technique similar to the sieve of Eratosthenes will state which of them are divisible by which primes p.

For this factoring algorithm, our factor base will consist of the prime two, and small odd primes p such that $(N|p) = 1$. We can create a table which, for each index x, contains the value of $(M + x)^2 - N$, and then we can divide all the even elements (every other element is even, so once we know where to start, we just consider alternate elements) by two. For each of the odd primes p, we just divide the elements of the appropriate arithmetic progressions by p. Of course, it is possible that the values of $(M + x)^2 - N$ are divisible by powers of p, and it would not be particularly expensive to perform trial division, since we need only consider those p which we know to divide $(M + x)^2 - N$. Alternatively, we can consider for which values of x the congruence $(M + x)^2 \equiv N$ (mod p^2) is soluble, and deduce additional arithmetic progressions in which we know that every value of $(M + x)^2 - N$ is divisble by p^2, and so on. This method can also be adapted for computers where division is a slow operation: rather than storing $(M + x)^2 - N$ and dividing it by p, we can store $\log((M + x)^2 - N)$ and subtract $\log p$ from it. This is particularly appropriate when factoring large numbers, as a sufficiently accurate approximation to $\log((M+x)^2-N)$ can be stored in a single computer word even when $(M + x)^2 - N$ requires several words to store it.

There are several important variations on this algorithm. There is a "large prime variant" analogous to the large prime variant we described for the continued fraction algorithm. Another variant, the "multiple polynomial quadratic sieve", uses several different polynomials instead of the one $Q(x)$, since these can be chosen to have more small values than $Q(x)$ has. Both variants can be employed together, and were so used in finishing the factorizations announced at the beginning of section 1. The best versions of this algorithm have running time proportional to $L(N)$.

6. *The Diffie-Hellman cryptographic method*

How can two people exchange information secretly, with no pre-arranged secret key? This may seem impossible, but the following analogy explains how it can be done. Suppose that A wishes to send B a large sum of money. He knows that the carriers always deliver parcels, but that they have the unfortunate habit of opening them first and taking money, or copying any keys they find in them. He could send a locked box to B, which would be delivered, but then he has the problem of sending B the key. He could send the key in a locked box, but then he has the problem of sending the key to the box containing the key .... What he can do is send B a box secured with a padlock, the key to which he retains. B cannot open this box, but he can place his own padlock on it, and send the box back to A. A can then remove his padlock, and return the box to B, who can unlock the box and recover the money. The method is perfectly secure, since the box is locked whenever it is in transit.

How do we convert this idea into a useful computer-oriented encryption scheme? First, we represent the message to be transferred as a sequence of integers to the modulus N, where N is a publicly-agreed large integer. Then our problem is to transfer these integers, and if we can transfer one such integer, we can transfer several by repeating the procedure (in practice, we generally transfer one integer, which is the key to some other cryptographic scheme for transferring large messages efficiently).

One possible method for conveying the message x is the following. A and B each think of a random number, say a and b, which have to be relatively prime to N. Then the sequence of exchanges between A and B can be summarised as

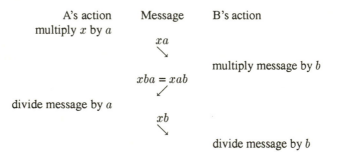

A's action	Message	B's action
multiply x by a	xa	
		multiply message by b
	$xba = xab$	
divide message by a	xb	
		divide message by b

where all multiplications and divisions take place to the modulus N, which is why we needed a and b to be relatively prime to N. The numbers a and b correspond to the two padlocks in the analogy given above, and the fact that multiplication is commutative, so that it does not matter in which order we multiply and divide by a and b, corresponds to the fact that the two padlocks can be added or removed in any order.

However, there is a serious flaw in this method. Consider the cryptanalyst who succeeds in obtaining all three messages. In isolation they tell him nothing, but if he has all three, he can compute

$$x = \frac{xa \times xb}{xab} \quad (\text{mod } N).$$

Strictly speaking, this will only work if x is relatively prime to N, since otherwise he will only obtain x to the modulus $N/\gcd(N, x)$. But the chance of x having a large factor in common with N is very small, and he will obtain "nearly all" the message. He could compute a or b as xab/xa or xab/xb, and then try all possibilities for a (mod N) (or b (mod N)), knowing a (mod $N/\gcd(N, x)$) (or b (mod $N/\gcd(N, x)$)), to see which gave sensible values for x. In practice this is not a difficulty, and the cryptanalyst can decipher these messages easily.

Hence we need a less vulnerable protocol for exchanging these digits: the one we shall give is the one Diffie and Hellman originally proposed. Instead of relying on multiplication and division, we shall rely on exponentiation and the extraction of roots. We shall consider this to a prime modulus P, rather than a general modulus N, though other choices are possible. We recall from III.2 that, if k is relatively prime to $P - 1$, then every number has a unique k-th root to the modulus P. This can be computed by finding a number l such that $kl \equiv 1$ (mod $P - 1$), and then the calculation of l-th powers is equivalent to the calculation of k-th roots. So now let A and B choose numbers a and b relatively prime to $P - 1$, and engage in the following dialogue

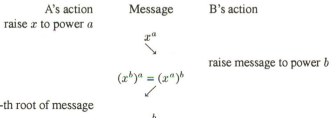

A's action   Message   B's action

raise x to power a

x^a

raise message to power b

$(x^b)^a = (x^a)^b$

take a-th root of message

x^b

take b-th root of message

where all calculations take place to the modulus P. a and b can be chosen to be large, in view of the efficient methods of raising to powers described in section 2.

Now what does the cryptanalyst do? The wise cryptanalyst re-reads the theory of III.2, where the concept of an *index* was introduced (except that cryptanalysts tend to use the term *discrete logarithm* rather than index). Let

ρ be any primitive root to the modulus P, then the index of any (non-zero) element x is that number ξ such that $\rho^{\xi} = x$. The index of x^a is then $a\xi$ (mod $P - 1$). The exchange above, when viewed as an exchange of indices, looks like this:

	A's action	Index of Message	B's action

A's action   Index of   B's action

           Message

raise x to power a

$$a\xi$$
$$\searrow$$

                     raise message to power b

$$ab\xi = ba\xi$$
$$\nearrow$$

take a-th root of message

$$b\xi$$
$$\searrow$$

                        take b-th root of message

and our cryptanalyst is back on familiar territory. Unless ξ has a factor in common with $P - 1$, he can determine ξ, and hence x, exactly. If there is such a common factor, he can still determine a to the modulus $(P - 1)/\gcd(P - 1, \xi)$ and then try all consistent values of a (mod $(P - 1)$) to find one that gives plausible values of x. The only trouble is that the cryptanalyst has to compute two or three indices, and the methods of III.2 are not efficient for large values of P. The most efficient methods currently known for finding indices to the modulus P have a running time proportional to some power of $L(P)$, which depends on P in the same way as the factoring algorithms described in the previous section.

7. The RSA cryptographic method

The basic purpose of this method, which is named after its inventors Rivest, Shamir and Adleman, is to provide a *one-way* method of secure communication. This is not as restrictive as it might seem, since a two-way secure method can be constructed trivially from two one-way secure methods, one in each direction. Also, a one-way method can be used to send a key for a more efficient cryptosystem for two-way communication. Let us suppose that person A wishes to enable other people to send him secure messages, which cannot be deciphered by those who manage to read them. A selects two distinct prime numbers P and Q, which must be sufficiently large that no adversary could factor $N = PQ$ except by luck. This may well mean that P and Q have to have over 100 digits each, and certainly means that P and Q should not be too close together, otherwise Fermat's method (I.9) may be used to factor N. A then chooses a number x relatively prime to $\phi(N) = (P - 1)(Q - 1)$ and publishes (one can think of a message in the personal columns of a newspaper) the values of N and x.

Anyone wishing to send a message to A then divides it up into digits to the base N (taking care to avoid extremely small digits) and transmits each digit a by sending $a^x \pmod{N}$ (which is computed by the repeated squaring method of section 2). A has to decode this message by computing the x-th roots of the digits received — these are unique since x is relatively prime to $\phi(N)$. By applying Euclid's algorithm to x and $\phi(N)$, A can compute an x' such that $xx' \equiv 1 \pmod{\phi(N)}$, as in the previous section. Raising to the x'-th power is then the same as taking x-th roots. In practice, A computes x' as soon as x has been chosen, and then forgets about P and Q.

Obviously, anyone who can factorize N can repeat A's calculation of x', and hence crack the code. So cracking this code is no harder than factorizing N. Suppose now that some-one knows x' such that $xx' \equiv 1 \pmod{\phi(N)}$, so that he can crack the code. Then that person can compute $xx' - 1 = M\phi(N)$ for some apparently unknown M. But $\phi(N)$ is a number slightly smaller than N, so M is slightly larger than $(xx' - 1)/N$, and computing this quotient and rounding it up will determine M. Once M is known, $\phi(N)$ is known, and $N + 1 - \phi(N)$ is $P + Q$. If we call R the value of $P + Q$, then the code-breaker knows $N = PQ = (R - Q)Q$, and Q is one of the roots of the quadratic equation $Q^2 - RQ + N = 0$, and P is the other root.

We have shown that a knowledge of the original x' leads to a factorization of N. However, the code-breaker does not necessarily have to find the original x'. Though no such way is currently known, there might be a method for taking x-th roots that did not rely on exponentiation at all. Even if exponentiation is used, the code-breaker might not have to know the original x' at all — any x'' such that $xx'' \equiv 1 \pmod{\hat{\phi}(N)}$ will do. If $\gcd(P - 1, Q - 1)$ is small, then the techniques of the previous paragraph can be adapted to find the factorization of N. If $\gcd(P - 1, Q - 1)$ is large, then we may be able to use other methods to factor N.

NOTES

Some of the earliest uses of electronic computers were in the search for large prime numbers: J.C.P. Miller and D.J. Wheeler found the prime $p = 180(2^{127} - 1)^2 + 1$, whose expanded form has 79 digits in 1951 (see Nature **168** p. 838). They proved that it was prime by exhibiting an x such that $x^{p-1} \equiv 1 \pmod{p}$ and $x^{(p-1)/d} \not\equiv 1 \pmod{p}$ for all prime divisors d of $p - 1$, viz. $d = 2, 3, 5, 2^{127} - 1$.

§1. The first two factorizations mentioned here were announced jointly by Mark Manasse of the Digital Equipment Corporation's Systems Research Center and Arjen

Lenstra of Bell Communications Research, on 26th April 1990 and 4th January 1991 respectively. The method used is known as "ppmpqs": the double-partial multiple-polynomial quadratic sieve, a development of the methods explained in section 5. For the factorization of the 116-digit factor of $10^{142} + 1$, it is estimated that some 600 computers throughout the world, contributing the equivalent of a one million instructions per second computer working for 400 years, worked on generating a set of 142,000 linear equations, which, using a very advanced method, were then solved on a parallel computer system. The factor 380623849488714809 of $10^{142} + 1$ had been found in 1986 by Harvey Dubner, using the elliptic curve algorithm (see section 5). The third factorisation was announced by Herman te Riele of the Centrum voor Wiskunde en Informatica in Amsterdam on the 11th February 1991. The 101-digit product of the last two factors holds, at the moment of writing, the record of most difficult number factored on a single computer. The sieving process took 475 hours, and the linear equation solving about half an hour, on a Cray Y-MP4/464. While this edition was being prepared, Mark Manasse announced the factorisation of $10^{151} - 1$, where a 113-digit number was factored as the product of 56 and 58-digit primes, of $2^{499} + 1$, where a 104-digit number was factored as the product of 52 and 53-digit primes, and of $2^{461} + 1$, where a 102-digit number was factored as the product of 37 and 66-digit primes.

The definitive reference for the best way of implementing long division etc. is Knuth's encyclopaedic *The Art of Computer Programming II: Semi-Numerical Algorithms* (Addison-Wesley, 1981). This also contains descriptions of the various faster algorithms of computer science, a lengthy treatise on random numbers, which treats the statistical as well as the arithmetical properties of these sequences, and also descriptions of Pollard's and Rabin's algorithms.

A simple application of congruences to hash tables can be found in a paper by F.R.A. Hopgood and J.H. Davenport called *The Quadratic Hash Method when the table size is a power of 2*, which appeared in the Computer Journal **15** (1973) pp. 314–315.

§2. D.H. Lehmer's proofs of the Lucas-Lehmer tests appeared in Annals of Mathematics (2) **31** (1930) pp. 419–448 and J. London Math. Soc. **10** (1935) pp. 162–165. To test whether $N = 2^p - 1$ is prime, we first check that p is prime, then construct the sequence $r_1 = 4, r_2 = 14, \ldots, r_{i+1} \equiv r_i^2 \pmod{N}$ and check that $r_{p-1} \equiv 0 \pmod{N}$.

Carmichael's original paper *On Composite Numbers P which satisfy the Fermat Congruence $a^{P-1} \equiv 1 \pmod{P}$* appeared in Amer. Math. Monthly 19(1912) pp. 22–27. Carmichael numbers have been intensively investigated: see the paper *The Pseudoprimes to $25 \cdot 10^9$* by Pomerance, Selfridge and Wagstaff, which appeared in Math. Comp. **35** (1980) pp. 1003–1026. This paper shows that there are 2163 Carmichael numbers less than 25×10^9. It is unknown whether there are infinitely many Carmichael numbers, though most experts believe that there are.

Rabin's original paper, called *Probabilistic Algorithm for Testing Primality*, appeared in J. Number Theory **12** (1980) pp. 128–138. The subject of alternative certificates is an active research area: one recent paper is by S. Goldwasser and J. Killian *Almost All Primes Can be Quickly Certified*, which appeared in the Proceedings of the 1986 Symposium on the Theory of Computing. Using the Elliptic Curve Primality Proving method due to A.O.L. Atkin, F. Morain proved the primality of the 1065-digit number $(2^{3539} + 1)/3$ using a month and a half of (Sun 3/60) computer time — see INRIA Research Reports 1152 and 1256. We note that a single application of Rabin's method takes about just over two hours on a similar machine, so we have to pay dearly for the certainty of a certificate.

§3. J. von Neumann, one of the very early pioneers of digital computing, seems to have suggested the mid-square method in about 1946. The linear congruential method was introduced by D.H. Lehmer in 1949. Knuth's book is the best source of criteria for random number generators: our arithmetic criteria are identical to his. The serious user of such sequences should use the various statistical tests described by Knuth. The values for $n = 2^{32}$ were supplied by Mr. N.M. Maclaren of the University of Cambridge Computing Service.

§4. Pollard's original paper, called *A Monte Carlo Method for Factorization*, appeared in B.I.T. **15** (1975) pp. 331–334. There have since been many minor improvements to it, but the outline given in the present book conveys the general principles. Some improvements are described by Montgomery in *Speeding the Pollard and Elliptic Curve Methods of Factorization* which appeared in Math. Comp. **48** (1987) pp. 243–264.

§5. The elliptic curve factoring method is described by H.W. Lenstra, Jr., in *Factoring Integers with Elliptic Curves*, which appeared in Annals of Mathematics (2nd. Ser.) **126** (1987) pp. 649–673. The statement we quote, that the running time is proportional to $L^2(p) \log^2 n$, depends on p being large, is only true "on average", since there is a remote possibility that the factor we find is actually n itself, and has only been proved if we assume a (very plausible) conjecture on the distribution of the numbers obtained during the runnig of the algorithm.

The use of multiple congruences of the form "$A^2 \equiv$ product of small primes" to factor numbers seems to be due to Kraitchik, who published it in his *Recherches sur la Théorie des Nombres Tome II: Factorisation*, Gauthier-Villars, Paris, 1929. The use of continued fractions to generate the congruences is due to Lehmer and Powers' paper *On Factoring Large Numbers* Bull. A.M.S. **37** (1931) pp. 770–776. Knuth gives a very elegant formulation of the continued fraction algorithm on pp. 381–2, and applies it to 197209, as we have done.

The quadratic sieve method of generating these congruences is due to Pomerance, and described in his paper *The Quadratic Sieve Factoring Algorithm*, which appeared in Proc. EUROCRYPT '84 (Springer Lecture Notes in Computer Science 209, ed. T. Beth, N. Cot & I. Ingemarsson, Springer-Verlag, Berlin-Heidelberg-New York, 1985) pp. 169–182. A recent survey of these methods is given by Wagstaff and Smith's paper *Methods of Factoring Large Integers* in Number Theory New York 1984–85 (Springer Lecture Notes in Mathematics 1240, ed. D.V. Chudnovsky, G.V. Chudnovsky, H. Cohn & M.B. Nathanson, Springer-Verlag, Berlin-Heidelberg-New York, 1987) pp. 281–303. The "multiple polynomial" variation is described by Silverman in *The Multiple Polynomial Quadratic Sieve* which appeared in Math. Comp. **48** (1987) pp. 329–339.

§6. The original Diffie–Hellman paper, *New Directions in Cryptography*, appeared in IEEE Trans. Inform. Theory **IT-22** (1976) pp. 644–654, and the method is also described in U.S. patent number 4,200,770. There has been much work recently on advanced methods for computing indices. A good description of several of these methods is given in the article *Discrete Logarithms in $GF(p)$* by Coppersmith, Odlyzko and Schroeppel, which appeared in Algorithmica **1** (1986) pp. 1–15. If p doesn't have any particularly helpful properties (in particular if $p - 1$ has a very large prime factor) then the running time of the best algorithm they mention is roughly proportional to $L(p)$. A variant on the Diffie-Hellman scheme, which uses finite fields of size $N = 2^n$ rather than prime size, can be cracked in time proportional to $\exp(c \log^{1/3} N \log \log^{2/3} N)$ — see Coppersmith's paper *Fast Evaluation of Logarithms in Fields of Characteristic Two*, which appeared in IEEE Trans. Inform. Theory **IT-30** (1984) pp. 587–594.

§7. The original Rivest, Shamir and Adleman paper, *A method for obtaining digital signatures and public key cryptosystems* appeared in Communications of the ACM **21** (1978) pp. 120–126. This is also described in U.S. patent number 4,405,829.

EXERCISES

The marks [H] and [A] affixed to questions indicate that the questions are provided with hints and answers respectively. If both are provided [H, A], try the hint first. The mark [M] affixed to a question indicates that it requires a little more mathematical knowledge than was assumed in the body of the book, for example, elementary complex numbers or trigonometry. Although such matters are hard to judge, the mark [+] has been used to indicate questions, or parts of questions, that are thought to be somewhat harder than average. The first digit of a question number indicates which chapter it refers to.

1.01. Prove, by induction or otherwise, that:
   (a) the sum of the first n numbers is $n(n+1)/2$ (this result is commonly said to have been discovered by Gauss at a very early age: see, for example, E. T. Bell, *Men of Mathematics*, Simon & Schuster, New York, 1937 (reprinted Penguin, 1965));
   (b) the sum of their squares is $n(n+1)(2n+1)/6$;
   (c) the sum of their cubes is $n^2(n+1)^2/4$.

1.02. Define the *Fibonacci numbers*, F_n, by $F_1 = F_2 = 1$, and $F_n = F_{n-1} + F_{n-2}$ for $n > 2$. Prove, by induction or otherwise, that:
   (a) $F_n < \tau^n$, where τ is the *golden ratio* $(1 + \sqrt{5})/2$;
   (b) $F_n = (\tau^n - \sigma^n)/\sqrt{5}$, where $\sigma = -1/\tau = (1 - \sqrt{5})/2$.

1.03. Express each of the following numbers as the product of prime factors:
   999, 1001, 1729, 11,111 [+], 65,536, 6,469,693,230.
   [A]

1.04. Find five consecutive composite numbers. Find thirteen such numbers. Find ninety-nine such numbers. [A]

1.05. Evaluate n^2+n+41 for $n = 0, 1, 2, \ldots$. Does this formula (attributed to Euler) always give prime numbers? (41 is, in fact, the largest number that can be placed in Euler's formula: this is intimately connected with the fact that $163 = 4 \times 41 - 1$ is the largest number with $C(-d) = 1$ (see VI.7 or Shanks, *Proc. Symp. Pure Maths.*, 20 (American Mathematical Society, 1971), 415–40).) [A]

1.06. *Factorial n*, written $n!$, is the product $1 \times 2 \times 3 \times \ldots \times n$ of the first n numbers. Express 22! as the product of prime factors. [A]

1.07. [M] Show that, if 2^a is the highest power of 2 which divides $n!$, then a lies between $n-1$ and $n - \lfloor \log_2 (n+1) \rfloor$, where $\log_2$ is the conventional logarithm to the base 2, and $\lfloor x \rfloor$ is Knuth's *floor* symbol for the greatest integer not greater than x (also called the *integer part* of x), so that $\lfloor \log_2 (n+1) \rfloor$ is the exponent of the greatest power of 2 not greater than $n+1$. [H]

1.08. If $p \geq 5$ is prime, show that the sum of the products in pairs of the numbers $1, 2, \ldots, p-1$ is divisible by p. We do not count 1×1, and 1×2 precludes 2×1. [H]

1.09. [M] Consider 'integers' of the form $a+b\xi$, where a and b are ordinary integers, and ξ is undetermined, except that, when two integers are multiplied, ξ^2 is replaced by -5:

$$(a_1+b_1\xi)(a_2+b_2\xi) = (a_1 a_2 - 5 b_1 b_2) + (a_1 b_2 + a_2 b_1)\xi.$$

Show that the only *units* (divisors of 1) of the form $a+b\xi$ are $a = 1, b = 0$ and $a = -1, b = 0$, and define *prime number* in this system. Show that 2, 3, $1+\xi$, and $1-\xi$ are all primes, although $2 \times 3 = (1+\xi)(1-\xi)$.

Show also that it is not possible to find 'integers' x, y of this kind which satisfy the equation $3x - (1+\xi)y = 1$. [H]

1.10. [M, +] Show that the *Gaussian integers*, numbers of the form $a+bi$, where a and b are ordinary integers and $i^2 = -1$, have unique factorization. [H]

1.11. If $2^n - 1$ is prime, show that n is prime. Is the converse true? [A]

1.12. If $2^n + 1$ is prime, show that n is a power of 2. Is the converse true? [+] [A]

1.13. If P_1, P_2 are even perfect numbers with $6 < P_1 < P_2$, show that $P_2 > 16 P_1$.

1.14. If p, q are odd primes, show that $p^a q^b$ can not be perfect.

1.15. Show that, if c is any common factor of a and b, then

$$(a/c, b/c) = (a, b)/c,$$

where we use (a, b) to denote the highest common factor of a and b.

Show also that, if a and b both divide n, and are *coprime* (that is $(a, b) = 1$), then ab divides n.

1.16. How many divisors of 720 are there? What is their sum? [A]

1.17. Show that 120 is a *multiply perfect number*, that is that $\sigma(n) = kn$ for some $k > 2$. Can you find an example with $k > 3$? [A]

1.18. [+] We define a *balanced number* to be one whose average size of divisor, $\sigma(n)/d(n)$, is equal to $n/2$. Show that 6 is the only balanced number. [H]

1.19. Use the Euclidean algorithm to find the highest common factor of 18,564 and 30,030. Check your answer by writing each number as the product of prime powers. What is the least common multiple of these numbers? [A]

1.20. Find a formula for all pairs of integers x and y such that
$$113x - 355y = 1. \quad [A]$$

1.21. Factor 2501 by Fermat's difference of squares method. [A]

1.22. Use Captain Draim's algorithm to factor 1037. [A]

1.23. Show that the *binomial coefficient* $p!/r!(p-r)!$ is divisible by p if p is prime and $1 \leqq r < p$.

1.24. Prove that there are infinitely many primes of the form $6k - 1$.

2.01. Show that, if $a \equiv b \pmod{2n}$, then $a^2 \equiv b^2 \pmod{4n}$. More generally, show that, if $a \equiv b \pmod{kn}$, then $a^k \equiv b^k \pmod{k^2 n}$.

2.02. What numbers leave remainders 2, 3, 4, 5 respectively when divided by 3, 4, 5, 6? [A]

2.03. What is the smallest positive integer which leaves remainders 1, 2, ..., 9 respectively when divided by 2, 3, ..., 10? [A]

2.04. Solve the congruence $97x \equiv 13 \pmod{105}$. [A]

2.05. Find the remainder when $(102^{73} + 55)^{37}$ is divided by 111. [H, A]

2.06. Show that, if $a^{p-1} \equiv 1 \pmod{p}$ for all a $(1 \leqq a < p)$, then p is prime. Show that $2^{p-1} \equiv 1 \pmod{p}$ is possible without p being prime.

[+] Show that $a^{p-1} \equiv 1 \pmod{p}$ for all a $(1 \leqslant a < p, (a, p) = 1)$ does not imply that p is prime. Show that $a^{p-1} \equiv 1 \pmod{p}$ and $a^d \not\equiv 1 \pmod{p}$ for any proper divisor d of $p-1$ does prove that p is prime. [A]

2.07. For what values of n is $\phi(n)$ odd? [A]

2.08. Find all values of n (less than 50, say) for which $\phi(n) = 2^a$. (These are the numbers of sides of regular polygons that can be constructed using only a straight-edge and compasses.) [A]

2.09. Define $a(n)$ as the number of solutions of $\phi(x) = n$. Make a table of $a(n)$ (for $1 \leqq n \leqq 10$, say). (Carmichael's conjecture, that $a(n)$ is never 1, has been verified for $n \leqslant 10^{10,000}$.)

2.10. For what values of n is $\phi(n) = n/3$? Find a value of n such that $\phi(n) < n/5$. [A]

2.11. Show that n is prime if, and only if,

$$\sigma(n) + \phi(n) = nd(n).$$

2.12. Prove that, if p is an odd prime, then $(p-2)! \equiv 1 \pmod{p}$, and that, if p is a prime greater than 3, $(p-3)! \equiv (p-1)/2 \pmod{p}$.

2.13. If p is an odd prime, and $a+b = p-1$, show that $a!b! + (-1)^a \equiv 0 \pmod{p}$.

2.14. Solve the congruence $x^2 \equiv -1$ (a) mod 5, (b) mod 25, (c) mod 125. [H, A]

2.15. Solve the congruence $x^2 \equiv 17 \pmod{128}$. [A]

2.16. Solve, or prove insoluble, each of the congruences $x^3 \equiv 3$, $x^3 \equiv 7$, $x^3 \equiv 11$, each mod 19.

2.17. Show that, if $(2a, m) = 1$, solving the congruence $ax^2 + bx + c \equiv 0 \pmod{m}$ can be reduced to solving a congruence of the form $x^2 \equiv q \pmod{m}$.

2.18. Verify the following divisibility tests. Separate the decimal digits of a number n into blocks of three:

$$n = b_k(1000)^k + \ldots + b_2(1000)^2 + b_1(1000) + b_0.$$

Sum alternate blocks, so that $E = b_0 + b_2 + b_4 + \ldots$ and $D = b_1 + b_3 + \ldots$. Then 3^a divides n if, and only if, it divides $E + D$ ($a = 1$, 2, 3); 37 divides n if, and only if, it divides $E + D$; each of 7, 11 and 13 divides n if, and only if, it divides $E - D$.

2.19. Show that every fourth Fibonacci number (see question 1.02) is divisible by 3, that every fifth is divisible by 5, every sixth by 8 and every seventh by 13.

2.20. If $d = (a, b)$, show that $\phi(d)\,\phi(ab) = d\phi(a)\,\phi(b)$.

2.21. Show that if d divides n, $\phi(d)$ divides $\phi(n)$.

2.22. Show that every prime except 2 and 5 divides infinitely many numbers of the form 11, 111, 1111, 11111, ...

2.23. Solve the simultaneous congruences $x \equiv 3 \pmod 9$, $x \equiv 5 \pmod{10}$, $x \equiv 7 \pmod{11}$. [A]

2.24. Solve the simultaneous congruences $9y \equiv 3 \pmod{15}$, $5y \equiv 7 \pmod{21}$, $7y \equiv 4 \pmod{13}$. [A]

2.25. Solve the simultaneous congruences $z \equiv 2 \pmod{15}$, $z \equiv 7 \pmod{10}$, $z \equiv 5 \pmod 6$. [A]

3.01. Find the quadratic, cubic and fifth power residues, mod 7. **[A]**

3.02. Find the quadratic, cubic and fifth power residues, mod 11. **[A]**

3.03. Find the quadratic, fourth power, eighth power and sixteenth power residues, mod 17. **[A]**

3.04. Find the primitive roots mod each of the primes 3, 5, 7, 11, 13, 17 and 19. **[A]**

3.05. Show that 10 and 2 are solutions of $x^8 \equiv 1$, $x^9 \equiv 1$ (mod 73) respectively, and hence that 20 is a primitive root mod 73.

3.06. Show that 2^k has no primitive roots if $k > 2$.

3.07. Find all the primitive roots mod 27. **[A]**

3.08. Find all the primitive roots mod 125. **[A]**

3.09. Show that any primitive root mod p is, in the notation of equation (2), the product of numbers x_i of order $q_i^{a_i}$. **[H]**

3.10. Show that there are always $\phi(p-1)$ primitive roots mod p, where p is prime. Hence prove the remark at the bottom of p. 62, that the numbers there constructed are all different.

3.11. Show that the product of the primitive roots mod a prime $p > 3$ is congruent to 1 (mod 3). **[H]**

3.12. If $p = 4k+1$ is a prime and g is a primitive root mod p, show that $p-g$ is also a primitive root mod p.

3.13. Show that, if $p = 4k-1$ and g is a primitive root mod p, then $p-g$ is not.

3.14. If g is a primitive root mod p^2, prove that it is also a primitive root mod p. Is the converse true? **[A]**

3.15. If p and $4p+1$ are both primes, show that 2 is a primitive root mod $4p+1$. **[A]**

3.16. If $4k+1$ and $8k+3$ are both primes, show that 2 is a primitive root mod $8k+3$.

3.17. If $4k+3$ and $8k+7$ are both primes, show that -2 is a primitive root mod $8k+7$.

3.18. Construct a table of indices for the prime 41, using the primitive root 6. Check that, for each a, the indices for $\pm a$ differ by 20. **[A]**

3.19. Show that a square is congruent to 0, 1 or 4 (mod 8), and that a fourth power is congruent to 0 or 1 (mod 16).

3.20. Make a list of quadratic residues for each prime p, $3 \leqq p \leqq 19$. **[A]**

3.21. Find all sets of two decimal digits which can occur as the last two digits of a perfect square. **[A]**

3.22. Use Gauss's lemma to show that -2 is a quadratic residue of primes of the form $8k+1$ and $8k+3$, and a non-residue of primes of the form $8k+5$ and $8k+7$.

3.23. Use Gauss's lemma to show that 5 is a quadratic residue of primes of the form $10k \pm 1$, and a non-residue of those of the form $10k \pm 3$.

3.24. Which primes have -3 as a quadratic residue? [A]

3.25. Calculate the Legendre symbols $(-26|73)$, $(19|73)$ and $(33|73)$. [A]

3.26. Which of the following congruences are soluble?

(a) $x^2 \equiv 125 \pmod{1016}$;
(b) $x^2 \equiv 129 \pmod{1016}$;
(c) $x^2 \equiv 41 \pmod{79}$;
(d) $41x^2 \equiv 43 \pmod{79}$;
(e) $43x^2 \equiv 47 \pmod{79}$;
(f) $x^2 \equiv 151 \pmod{840}$.
[A]

4.01. Express 105/143, 112/153, 89/144 and 169/239 as continued fractions. [A]

4.02. Calculate $[3, 1, 4, 1, 6]$ and $[6, 1, 4, 1, 3]$. [A]

4.03. Write down the convergents to each of the following continued fractions:

(a) $1 + \dfrac{1}{1+} \dfrac{1}{1+} \dfrac{1}{1+} \dfrac{1}{1+} \dfrac{1}{1+} \dfrac{1}{1}$;

(b) $2 + \dfrac{1}{2+} \dfrac{1}{2+} \dfrac{1}{2+} \dfrac{1}{2+} \dfrac{1}{2+} \dfrac{1}{2}$;

(c) $2 + \dfrac{1}{4+} \dfrac{1}{4+} \dfrac{1}{4+} \dfrac{1}{4}$;

(d) $1 + \dfrac{1}{1+} \dfrac{1}{2+} \dfrac{1}{1+} \dfrac{1}{2+} \dfrac{1}{1+} \dfrac{1}{2}$.
[A]

4.04. Express each of the convergents from the previous question as a decimal fraction. [A]

4.05. Find the general solution in integers for each of the equations $355x - 113y = 1$ and $355x + 113y = 1$. [A]

4.06. Find the periodic continued fractions for $\sqrt{51}$ and $\sqrt{52}$. [A]

4.07. Show that the continued fraction for $\sqrt{n^2+1}$ is $n, \overline{2n}$.

4.08. Show that the continued fraction for $\sqrt{n(n+1)}$ is $n, \overline{2, 2n}$.

4.09. Choose a convergent to each of the continued fractions of question 4.03 (continuing the patterns if necessary) with a sufficiently large denominator to give approximations, correct to four decimal places, to $(1+\sqrt{5})/2$, $1+\sqrt{2}$, $\sqrt{5}$ and $\sqrt{3}$. [A]

4.10. Show that the quadratic irrational number $(4+\sqrt{37})/7$ is reduced, and find its purely periodic continued fraction. [A]

4.11. Find the first few partial quotients in the continued fraction for $\sqrt[3]{3}$. Give the corresponding convergents, and express them as decimal fractions. [A]

4.12. Write down the first few convergents to the continued fraction for e:

$$2+\cfrac{1}{1+}\cfrac{1}{2+}\cfrac{1}{1+}\cfrac{1}{1+}\cfrac{1}{4+}\cfrac{1}{1+}\cfrac{1}{1+}\cfrac{1}{6+}\cfrac{1}{1+}\cfrac{1}{1+}\cfrac{1}{8+}\cdots.$$

Which is the earliest continued fraction to approximate e to 6 decimal places? ($e \approx 2.718281828459045....$) [A]

4.13. Use alternate convergents to the continued fraction for $\sqrt{2}$ to give solutions of Pell's equations $x^2-2y^2 = 1$ and $x^2-2y^2 = -1$. Show that the numerators and denominators each satisfy the recurrence relation $u_{n+1} = 6u_n - u_{n-1}$.

4.14. In a similar way, relate the convergents to $\sqrt{3}$ with solutions to $x^2-3y^2 = 1$ and $x^2-3y^2 = -2$, and the recurrence relation $u_{n+1} = 4u_n - u_{n-1}$.

4.15. In a similar way, relate the convergents to $\sqrt{5}$ with solutions to $x^2-5y^2 = 1$ and $x^2-5y^2 = -1$, and the recurrence relation $u_{n+1} = 18u_n - u_{n-1}$.

4.16. N is said to be *square* if $N = m^2$, and N is said to be *triangular* if $N = n(n+1)/2$. Find those numbers that are both square and triangular. [H, A]

5.01. Which of the following numbers can be expressed as the sum of two squares: 97, 221, 300, 490, 729, 1001? [A]

5.02. Verify that $(a^2+b^2)(c^2+d^2) = (ac+bd)^2+(ad-bc)^2 = (ac-bd)^2 +(ad+bc)^2$, and hence that, in general, such a product is expressible as the sum of two squares in at least two different ways. What is meant by 'in general' here? [A]

5.03. Use the above formula to show that a prime which is the sum of two squares can only be expressed in one way. [H, A]

5.04. Illustrate the proof that primes of the form $4k+1$ are representable as the sum of two squares with the prime 449 and the solution $z = 67$ of the congruence $z^2+1 \equiv 0 \pmod{449}$.

5.05. Illustrate Legendre's construction by showing that the appropriate complete quotient in the continued fraction for $\sqrt{449}$ is $(20+\sqrt{449})/7$. [H]

5.06. Illustrate Serret's construction by expanding $449/67$ as a continued fraction. [H]

5.07. Verify Euler's identity for $(a^2 + b^2 + c^2 + d^2)(A^2 + B^2 + C^2 + D^2)$.

5.08. Express 103 as the sum of four squares in several different ways. [A]

5.09. Find solutions to $x^2 \equiv 2$ and $y^2 \equiv -3 \pmod{103}$, put $x^2 + y^2 + 1 = 103m$ and deduce a representation of 103 as the sum of four squares.

5.10. Which of the following numbers can be expressed as the sum of three squares: 607, 307, 284, 568, 1136? [A]

5.11. Show that the number of numbers less than 2^{2k+1} which are not expressible as the sum of three squares is $(2^{2k} - 1)/3$.

6.01. Show that $13x^2 + 36xy + 25y^2$ and $58x^2 + 82xy + 29y^2$ are each equivalent to the form $x^2 + y^2$.

6.02. Prove that the forms $ax^2 \pm bxy + cy^2 \; (-a < b < a < c)$ are not (properly) equivalent if $b \neq 0$.

6.03. Verify that, if
$$ax^2 + bxy + cy^2 = AX^2 + BXY + CY^2,$$
where $x = pX + qY$ and $y = rX + sY$, then
$$B^2 - 4AC = (b^2 - 4ac)(ps - qr)^2.$$

6.04. Use operations (i) and (ii) on p. 141 to reduce the forms $(13, 36, 25)$ and $(58, 82, 29)$ of question 6.01 to the equivalent reduced form $(1, 0, 1)$.

6.05. What are the discriminants of the forms $199x^2 - 162xy + 33y^2$ and $35x^2 - 96xy + 66y^2$? Are these forms equivalent? [A]

6.06. Show that a prime p can be represented by the form $x^2 + 2y^2$ if, and only if, $p = 2$ or $p \equiv 1$ or $3 \pmod 8$. [H, A]

6.07. Show that a prime p can be represented by the form $x^2 + 3y^2$ if, and only if, $p = 3$ or $p \equiv 1 \pmod 6$. [H]

6.08. Show that 23 has -5 as a quadratic residue, but that 23 is not representable by the form $x^2 + 5y^2$. Is 46 so representable? Show that the following conditions are necessary (but not sufficient!) for $x^2 + 5y^2$ to be prime: $(x, y) = 1$, $x \not\equiv y \pmod 2$, $xy \equiv 0 \pmod 3$. [A]

6.09. Use Dirichlet's class-number formula to calculate the class number of the discriminant $-p$ for some of the primes given in Table 11, and check that this agrees with the number of forms listed.

6.10. [+] If ρ is the number of quadratic residues in $(1 \leqq r \leqq \frac{1}{2}(\rho-1))$, and ν is the number of non-residues, show that $C(-p) = \frac{1}{3}(\rho-\nu)$ for $p \equiv 3$ (mod 8). [A]

6.11. [+] With the same notation as the previous question, show that $C(-p) = \rho-\nu$, for $p \equiv 7$ (mod 8).

6.12. Verify that $C(-163) = 1$.

7.01. Find all integer right-angled triangles with one side of length 25. [A]

7.02. Show that it is impossible to draw an equilateral triangle with each of its vertices at lattice points (with integer coordinates). [H]

7.03. Find all solutions in integers of $x^2 = y^2 + 3z^2$. [A]

7.04. Find all solutions in integers of $x^2 + y^2 = 2z^2$. [H, A]

7.05. Find all solutions in integers of $x^2 + 2y^2 = 3z^2$. [A]

7.06. [M] Find all triangles ABC with integer sides and angle A twice angle B. [H, A]

7.07. [M, +] Find all integer triangles with one angle of $60°$. [A]

7.08. Show that the equations $2x^2 + 5y^2 = z^2$ and $3x^2 + 5y^2 = z^2$ have no solutions in integers other than $(0, 0, 0)$.

7.09. Find an infinite set of essentially different solutions to equation (12). [A]

8.01. Find a non-prime that passes Rabin's test for the "random" x-value 2 [+] [H] [A].

8.02. Produce a good linear congruential method for simulating throws of a die.

8.03. Produce a good linear congruential method for simulating throws of two dice.

8.04. Pollard's ρ method requires the computation of a greatest common divisor every step. Can the cost of this be reduced? [A].

HINTS

1.07. Consider the forms of n for which the difference $n - a$ is maximal, and those for which it is minimal.

1.08. Start with the sum of all possible products, and subtract the terms we do not want.

1.09. To show that $1 + \xi$ is a prime, we first define the *Norm* of an integer $a + b\xi$ to be $a^2 + 5b^2$. Then $\text{Norm}(xy) = \text{Norm}(x)\,\text{Norm}(y)$, for integers x, y of this form. $\text{Norm}(1 + \xi) = 6$, so any factors of $1 + \xi$ must have Norms dividing 6. But the elements of Norm 1 are the units, those of Norm 6 are $1 + \xi$ and $-1 - \xi$, and there are no elements of Norm 2 or 3.

1.10. Define the Norm of $a + bi$ to be $a^2 + b^2$. If we have two Gaussian integers $a + bi$ and $c + di$, then their quotient is a complex number, and the closest Gaussian integer to it is at most $\sqrt{2}/2$ away, that is there is a Gaussian integer $e + fi$ such that

$$\text{Norm}(e + fi - (a + bi)/(c + di)) \leq \sqrt{2}/2 < 1.$$

Hence

$$\text{Norm}((e + fi)(c + di) - (a + bi)) < \text{Norm}(c + di).$$

This equation is analogous to (2), and lets us define Euclid's algorithm for Gaussian integers. The proof of unique factorization then follows as for the ordinary integers.

1.18. If n is the product of primes p_i, then $\sigma(n) < n\Pi\, p_i/(p_i - 1)$.

2.05. Work mod 3 and mod 37, and then combine the results.

2.14. If we have found a solution x_0 to $x^2 \equiv -1 \pmod 5$, then we can write $x = x_0 + 5x_1$, and find x_1 so that $x^2 \equiv -1 \pmod{25}$, and then we write $x = x_0 + 5x_1 + 25x_2$. This process of finding solutions mod a high power of a prime by 'lifting' a solution from a lower power is termed *Hensel's lemma*.

3.09. Let $n_i = (p-1)/q_i^{a_i}$. Then, if g is our primitive root, g^{n_i} has order $q_i^{a_i}$. Since the n_i have no factor in common, there exist integers l_i such that the sum of the $l_i n_i$ is 1. Then take $g^{l_i n_i}$.

3.11. If g is a primitive root, so is $1/g$.

4.16. If $m^2 = n(n+1)/2$, then $8m^2 = (2n+1)^2 - 1$. Now use question 4.13.

5.03. If $p = P^2 + Q^2 = R^2 + S^2$, where we can choose P and R to be even and Q and S to be odd (excluding the case $p = 2$), then $(Q+S)(Q-S) = (R+P)(R-P)$.

5.05. $\sqrt{449} = 21, \overline{5, 3, 1, 1, 1, 7, 1, 5, 5, 1, 7, 1, 1, 1, 3, 5, 42}$.
We therefore want the complete quotient after
$$21, 5, 3, 1, 1, 1, 7, 1, 5.$$

5.06. $\dfrac{449}{67} = 6 + \dfrac{1}{1+} \dfrac{1}{2+} \dfrac{1}{1+} \dfrac{1}{6}$.
Hence $x = [6, 1, 2]$ and $y = [6, 1]$.

6.06. Congruences mod 8 show that only these primes can be represented. If $p \equiv 1$ or $3 \pmod 8$, then -2 is a quadratic residue, so the equation $\alpha^2 = -2 + \beta p$ is soluble.

6.07. -3 is a quadratic residue of primes of the form $6k+1$.

7.02. We may assume that one of the vertices is the origin, and that at least one of the other coordinates is odd (otherwise we consider the triangle half the size). Now take congruences mod 4.

7.04. For a primitive solution, x and y are both odd, so write $x = p+q$, $y = p-q$.

7.06. Use the sine rule.

8.01. If k is such that $k+1$ and $3k+1$ are both prime, let n be $(k+1)(3k+1)$. $\phi(n) = 3k^2$, $\dot{\phi}(n) = 3k$ and k certainly divides $n-1$. So if 2 happens to be a perfect cube root to the modulus n (in fact, to the modulus $3k+1$ suffices) then $2^k \equiv 1 \pmod n$. The only question then is whether Rabin's test is more rigorous than Fermat's, which depends on the quadratic character of 2 to the moduli $k+1$ and $3k+1$.

ANSWERS

1.03. $3^3 \times 37$, $7 \times 11 \times 13$, $7 \times 13 \times 19$, 41×271, 2^{16},
$2 \times 3 \times 5 \times 7 \times 11 \times 13 \times 17 \times 19 \times 23 \times 29$.

1.04. 24, 25, ..., 28; 114, 115, ..., 126; $100! + x$ for $2 \leqq x \leqq 100$ (though a range with smaller numbers almost certainly exists).

1.05. No: $n = 40$ gives 41^2, $n = 41$ gives 41×43.

1.06. $2^{19} \times 3^9 \times 5^4 \times 7^3 \times 11^2 \times 13 \times 17 \times 19$.

1.11. If $n = ab$, then
$$2^n - 1 = (2^b - 1)(1 + 2^b + 2^{2b} + \ldots + 2^{(a-1)b}).$$
The converse is not true: $2^{11} - 1 = 23 \times 89$.

1.12. If p is an odd prime factor of n, so that $n = mp$, then
$$2^n + 1 = (2^m + 1)(1 - 2^m + 2^{2m} - \ldots + 2^{(p-1)m}).$$
Fermat thought that the converse was true, but Euler discovered that $2^{32} + 1 = 641 \times 6,700,417$.

1.16. 30. 2418.

1.17. $\sigma(30,240) = 4 \times 30,240$ (due to Descartes). Examples have been found with $k = 8$ (see Guy, B. 2).

1.19. 546. $18,564 = 2^2 \times 3 \times 7 \times 13 \times 17$;
$30,030 = 2 \times 3 \times 5 \times 7 \times 11 \times 13$. $1,021,020$.

1.20. $x = 22 + 355t$, $y = 7 + 113t$, where t is any integer.

1.21. 41×61.

1.22. 17×61.

2.02. $60k + 59$, where k is any integer.

2.03. 2519.

2.04. $x \equiv 64 \pmod{105}$.

2.05. 46.

2.06. If $a^{p-1} \equiv 1 \pmod{p}$, then a^{p-1} and p are coprime, and therefore a and p are coprime. If this is true for all a ($1 \leqslant a < p$) then p is prime.
   $2^{340} \equiv 1 \pmod{341}$.
   $a^{560} \equiv 1 \pmod{561}$ for all a coprime to 561, since $a^2 \equiv 1 \pmod{3}$, $a^{10} \equiv 1 \pmod{11}$ and $a^{16} \equiv 1 \pmod{17}$. If $a^d \not\equiv 1 \pmod{p}$ for any proper divisor d of $p-1$, it follows that the values $a, a^2, \ldots, a^{p-1}$ are all distinct $\pmod{p}$, and therefore that they take all values between 1 and $p-1$ in some order. But a^k is coprime to p, and therefore all the numbers between 1 and $p-1$ are coprime to p, so p is prime.

2.07. $\phi(1) = \phi(2) = 1$; otherwise $\phi(n)$ is even.

2.08. 3, 4, 5, 6, 8, 10, 12, 15, 16, 17, 20, 24, 30, 32, 34, 40, 48, ...

2.10. $n = 2^a 3^b$, with $a > 0$, $b > 0$.   30,030.

2.14. $x \equiv \pm 2 \pmod{5}$, $x \equiv \pm 7 \pmod{25}$, $x \equiv \pm 57 \pmod{125}$.

2.15. $x \equiv \pm 23$ or $\pm 41 \pmod{128}$.

2.23. $x \equiv -15 \pmod{990}$.

2.24. $y \equiv -343 \pmod{1365}$.

2.25. $z \equiv -13 \pmod{30}$.

3.01. $\{1, 2, 4\}$, $\{\pm 1\}$, $\{\text{all}\}$.

3.02. $\{1, -2, 3, 4, 5\}$, $\{\text{all}\}$, $\{\pm 1\}$.

3.03. $\{\pm 1, \pm 2, \pm 4, \pm 8\}$, $\{\pm 1, \pm 4\}$, $\{\pm 1\}$, $\{1\}$.

3.04. $\{2\}$, $\{\pm 2\}$, $\{-2, 3\}$, $\{2, -3, -4, -5\}$, $\{\pm 2, \pm 6\}$, $\{\pm 3, \pm 5, \pm 6, \pm 7\}$, $\{2, 3, -4, -5, -6, -9\}$.

3.07. 2 and 5 $\pmod{9}$.

3.08. $\pm 2, \pm 3, \pm 8, \pm 12 \pmod{25}$.

3.14. No: 7 is a primitive root mod 5, but not mod 25.

3.15. $p \neq 2$, therefore p is odd and $4p+1 = 8k+5$ for some k. Therefore 2 is not a quadratic residue mod p. Now, if 2 is not a primitive root mod $4p+1$, then either $2^4 \equiv 1 \pmod{4p+1}$ or $2^{2p} \equiv 1 \pmod{4p+1}$. The first is clearly impossible, and the second implies that 2 is a square.

3.18.

a	1	2	3	4	5	6	7	8	9	10
ind	40	26	15	12	22	1	39	38	30	8
$-a$	40	39	38	37	36	35	34	33	32	31
ind	20	6	35	32	2	21	19	18	10	28
a	11	12	13	14	15	16	17	18	19	20
ind	3	27	31	25	37	24	33	16	9	34
$-a$	30	29	28	27	26	25	24	23	22	21
ind	23	7	11	5	17	4	13	36	39	14

3.20. $3: \{1\}$. $5: \{\pm 1\}$. $7: \{1, 2, -3\}$. $11: \{1, -2, 3, 4, 5\}$.
$13: \{\pm 1, \pm 3, \pm 4\}$. $17: \{\pm 1, \pm 2, \pm 4, \pm 8\}$.
$19: \{1, -2, -3, 4, 5, 6, 7, -8, 9\}$.

3.21. 00, 25, $e1$, $e4$, $e9$ (where e is any even digit), $d6$ (where d is any odd digit).

3.24. $p = 6k + 1$.

3.25. $-1, 1, -1$.

3.26. (b), (d), (e).

4.01. $\dfrac{1}{1+}\dfrac{1}{2+}\dfrac{1}{1+}\dfrac{1}{3+}\dfrac{1}{4+}\dfrac{1}{2}$,

$\dfrac{1}{1+}\dfrac{1}{2+}\dfrac{1}{1+}\dfrac{1}{2+}\dfrac{1}{1+}\dfrac{1}{2+}\dfrac{1}{1+}\dfrac{1}{2}$,

$\dfrac{1}{1+}\dfrac{1}{1+}\dfrac{1}{1+}\dfrac{1}{1+}\dfrac{1}{1+}\dfrac{1}{1+}\dfrac{1}{1+}\dfrac{1}{1+}\dfrac{1}{1+}\dfrac{1}{2}$.

(Is it a coincidence that 89 and 144 are consecutive Fibonacci numbers?)

$\dfrac{1}{1+}\dfrac{1}{2+}\dfrac{1}{2+}\dfrac{1}{2+}\dfrac{1}{2+}\dfrac{1}{2+}\dfrac{1}{2}$.

4.02. 157 (for both).

4.03. (a) $\dfrac{1}{1}, \dfrac{2}{1}, \dfrac{3}{2}, \dfrac{5}{3}, \dfrac{8}{5}, \dfrac{13}{8}, \dfrac{21}{13}$;

(b) $\dfrac{2}{1}, \dfrac{5}{2}, \dfrac{12}{5}, \dfrac{29}{12}, \dfrac{70}{29}, \dfrac{169}{70}, \dfrac{408}{169}$;

(c) $\dfrac{2}{1}, \dfrac{9}{4}, \dfrac{38}{17}, \dfrac{161}{72}, \dfrac{682}{305}$;

(d) $\dfrac{1}{1}, \dfrac{2}{1}, \dfrac{5}{3}, \dfrac{7}{4}, \dfrac{19}{11}, \dfrac{26}{15}, \dfrac{71}{41}$.

4.04. (*a*) 1.0, 2.0, 1.5, 1.666..., 1.6, 1.625, 1.614...;
  (*b*) 2.0, 2.5, 2.4, 2.416..., 2.4137..., 2.41428..., 2.414201...;
  (*c*) 2.0, 2.25, 2.235..., 2.23611..., 2.23606...;
  (*d*) 1.0, 2.0, 1.666..., 1.75, 1.727..., 1.7333, 1.7317....

4.05. $x = -7 + 113t$, $y = -22 + 355t$ and
  $x = -7 + 113t$, $y = 22 - 355t$.

4.06. 7, $\overline{7, 14}$ and 7, $\overline{4, 1, 2, 1, 2, 14}$.

4.09. (*a*) 144 > 100, so 233/144 ($\approx$ 1.61806) is accurate to four decimal places. In fact 144/89 ($\approx$ 1.61798) is also accurate to four decimal places. The true answer $\approx$ 1.61803.
  (*b*) 169 > 100, so 408/169 ($\approx$ 2.41420) is accurate to four decimal places. In fact 169/70 ($\approx$ 2.41429) is also accurate to four decimal places. The true answer $\approx$ 2.41421.
  (*c*) 305 > 100, so 682/305 ($\approx$ 2.23607) is accurate to four decimal places. In fact 161/72 ($\approx$ 2.23611) is also accurate to four decimal places. The true answer $\approx$ 2.23607.
  (*d*) 153 > 100, so 265/153 ($\approx$ 1.73203) is accurate to four decimal places. In fact 79/56 ($\approx$ 1.73214) is also accurate to four decimal places. The true answer $\approx$ 1.73205.

4.10. $\overline{1, 2, 3}$.

4.11. 1, 2, 3, 1, 4, 1, ... 1/1 = 1.0, 3/2 = 1.5, 10/7 = 1.428..., 13/9 = 1.444..., 62/43 = 1.4418..., 75/52 = 1.4423....

4.12. 2/1, 3/1, 8/3, 11/4, 19/7, 87/32, 106/39, 193/71, 1264/465, 1457/536, 2721/1001 = 2.718281.

4.16. Convergents 3/2, 17/12, 99/70, ... yield $(m, n) = (1, 1)$, (6, 8), (35, 49), ... and the numbers 1, 36, 1225, ...

5.01. $97 = 9^2 + 4^2$, $221 = 10^2 + 11^2$ or $14^2 + 5^2$, $490 = 21^2 + 7^2$, $729 = 27^2 + 0^2$.

5.02. $a \neq 0$, $b \neq 0$, $c \neq 0$, $d \neq 0$, $a^2 \neq b^2$, $c^2 \neq d^2$, $\{a^2, b^2\} \neq \{c^2, d^2\}$.

5.03. By unique factorization, we can write $R + P = 2ac$, $R - P = 2bd$, $Q + S = 2ad$, $Q - S = 2bc$. Then $P = ac - bd$, $Q = ad + bc$, $R = ac + bd$, $S = ad - bc$, and $p = (a^2 + b^2)(c^2 + d^2)$.

5.08. $10^2 + 1^2 + 1^2 + 1^2$, $9^2 + 3^2 + 3^2 + 2^2$, $7^2 + 7^2 + 2^2 + 1^2$, $7^2 + 6^2 + 3^2 + 3^2$, $7^2 + 5^2 + 5^2 + 2^2$.

5.10. $307 = 17^2 + 3^2 + 3^2 = 15^2 + 9^2 + 1^2$, $568 = 18^2 + 12^2 + 10^2$.

6.05. -24. No: the reduced forms are $x^2 + 6y^2$ and $2x^2 + 3y^2$ respectively.

6.06. Then the form $px^2 + 2\alpha xy + \beta y^2$ has discriminant -8, and so has to be equivalent to $x^2 + 2y^2$. But it also represents p, by choosing $x = 1$, $y = 0$.

6.08. Congruences mod 5 show that $23 \neq x^2 + 5y^2$. $46 = 1^2 + 5 \times 3^2$.

6.10. Let A be the sum of all the quadratic residues, and B the sum of all the non-residues. 2 is a non-residue, so, if x is a residue, $2x$ is a non-residue. There are ν residues greater than $\frac{1}{2}p$, so

$$2A = B + \nu p. \qquad A + B = \frac{1}{2}p(p-1) = (\nu + \rho)p.$$

Solving these equations shows that $(B - A)/p = \frac{1}{3}(\rho - \nu)$.

7.01. (15, 20, 25),   (25, 60, 65),   (7, 24, 25),   (25, 312, 313).

7.03. $x = \pm(r^2 + 3s^2)t$, $y = \pm(r^2 - 3s^2)t$, $z = \pm(r^2 + s^2)t$, where r and s are coprime positive integers and t is a positive integer (or half-integer if r and s are both odd).

7.04. $x = \pm(r^2 + 2rs - s^2)t$, $y = \pm(s^2 + 2rs - r^2)t$, $z = \pm(r^2 + s^2)t$, where r and s are coprime positive integers and t is a positive integer.

7.05. $x = \pm(r^2 + 6rs + 3s^2)t$, $y = \pm(r^2 - 3s^2)t$, $z = (r^2 + 2rs + 3s^2)t$, where r and s are coprime positive integers and t is a positive integer (or half-integer if r and s are both odd).

7.06. The sides a, b, c are opposite the angles θ, 2θ and $180 - 3\theta$ respectively. Then, by the sine rule,

$$\frac{a}{\sin \theta} = \frac{b}{\sin 2\theta} = \frac{c}{\sin (180 - 3\theta)} = \frac{c}{\sin 3\theta}.$$

Now $\sin 2\theta = 2 \sin \theta \cos \theta$ and $\sin 3\theta = \sin \theta (4 \cos^2 \theta - 1)$, so $\cos \theta = b/2a$, and $b^2 - a^2 = ca$. Let $a = p^2 q$, where q is square-free. Then we can write $b = pqr$, and we deduce $a = p^2 q$, $b = pqr$ and $c = q(r^2 - p^2)$. We can make this representation unique by demanding that p and r have no common factor.

7.07. $a = (r^2 - rs + s^2)t$, $b = (2r - s)st$, $c = r(2s - r)t$, $c_1 = (r^2 - s^2 t)$, where $0 < s \leq r < 2s$ and $t > 0$.

7.09. An infinite set of solutions to $X^2 + 31 Y^2 = Z^2$ is $Z = p^2 + 31q^2$, $Y = 2pq$, $X = p^2 - 31q^2$, where one of p, q is even and the other odd, and they have no common factors. So we can take

$$x = 3(p^2 - 31q^2), \; y = 40pq + (p^2 + 31q^2), \; z = 62pq + 20(p^2 + 31q^2).$$

8.01. On the lines of the hint, let $k = 10$, so $n = 341 = 11 \cdot 31$. 2 is a perfect cube to the modulus 31 ($2 \equiv 4^3 \equiv 7^3 \equiv 20^3$). Unfortunately $2^{85} \equiv 32$, but $2^{170} \equiv 1$, so this time 2 passes Fermat's test, and fails Rabin's.

The next useful case is $k = 36$, with $n = 4033 = 37 \cdot 109$. $n - 1 = 2^8 \cdot 63$, and $2^{63} \equiv 3521$, while $2^{126} \equiv -1$, so 4033 does pass Rabin's test for the value 2. The question of numbers that pass Rabin's test for specific starting values is discussed by Arnault (*Le Test de Primalité de Rabin-Miller: Un nombre composé qui le "passe"*— Report 61, Université de Poitiers Département de Mathématiques,

November 1991) and by Davenport (in press), who suggests that, if the test values x are *published*, the number of x used should be proportional to $\log n$.

8.04. Yes, we can accumulate several values of the form $x_t - x_{t+i-j}$, multiply them together (to the modulus n) and compute the greatest common divisor of this product and n. We *may* miss a factorization this way, since the product *might* be divisible by all the factors of n, but this is extremely unlikely, and if it does happen, we can go back and try each $x_t - x_{t+i-j}$ in turn. Many implementors of this algorithm accumulate 10 such values at a time.

Furthermore, if we know that n has no prime factors smaller than B, which in practice we will do, we can abandon any computation of a greatest common divisor as soon as one of the numbers involved is less than B.

BIBLIOGRAPHY

This list contains a selection of books on the theory of numbers in general. References to works on special branches of the subject will be found in the notes given at the end of each chapter.

ENGLISH

DICKSON, L. E., *Introduction to the Theory of Numbers* (Chicago University Press, 1929); *History of the Theory of Numbers* (Carnegie Institute, Washington: vol. I, 1919; vol. II, 1920; vol. III, 1923); *Modern Elementary Theory of Numbers* (Chicago University Press, 1939)

GELFOND, A. O., and LINNIK, JU. V., *Elementary Methods in Analytic Number Theory* (Rand McNally, Chicago, 1965)

GUY, RICHARD K., *Unsolved Problems in Number Theory* (Springer, 1981)

HARDY, G. H., and WRIGHT, E. M., *Introduction to the Theory of Numbers* (Clarendon Press, Oxford, 5th ed., 1979)

LEVEQUE, W. J., *Topics in Number Theory* (2 vols., Addison–Wesley, Reading, Mass., 1956)

LEVEQUE, W. J. Ed., *Studies in Number Theory* (MAA studies in mathematics, 6. Prentice Hall, 1969)

MATHEWS, G. B., *Theory of Numbers* (Deighton Bell, Cambridge, 1892; Part I only published)

NAGELL, T., *Introduction to Number Theory* (John Wiley, New York, 1951)

ORE, O., *Number Theory and its History* (McGraw-Hill, New York, 1948)

RADEMACHER, H., *Lectures on Elementary Number Theory* (Blaisdell Pub. Co., 1964)

SHANKS, D., *Solved and Unsolved Problems in Number Theory* (Spartan Books, Washington D.C., 1962; reprinted by Chelsea Publ. Co., New York, 1978)

SIERPINSKI, W., *Elementary Theory of Numbers* (P.W.N., Warsaw, 1964); *A Selection of Problems in the Theory of Numbers* (Pergamon Press, 1964)

USPENSKY, J. V., and HEASLET, M. A., *Elementary Number Theory* (McGraw-Hill, New York, 1939)

VINOGRADOV, I. M., *An Introduction to the Theory of Numbers*, translated H. Popova (London, 1955)

WEIL, ANDRÉ, *Number Theory for Beginners* (Springer, 1979)

FRENCH

CAHEN, E., *Théorie des nombres* (2 vols., Hermann, Paris, 1924)

GERMAN

BACHMANN, P., *Niedere Zahlentheorie* (Teubner, Leipzig: vol. I, 1902; vol. II, 1910)

BESSEL-HAGEN, E., *Zahlentheorie* (Pascals Repertorium, vol. I, part 3; Teubner, Leipzig, 1929)

DIRICHLET, P. G. L., *Vorlesungen über Zahlentheorie*, edited by R. Dedekind (Vieweg, Braunschweig; 4th ed., 1894)

HASSE, H., *Vorlesungen über Zahlentheorie* (Springer, Berlin, 1950)

LANDAU, E., *Vorlesungen über Zahlentheorie* (3 vols., Hirzel, Leipzig, 1927; reprinted by Chelsea, New York)

SCHOLZ, A., *Einführung in die Zahlentheorie* (Sammlung Göschen, no. 1131, de Gruyter, Berlin, 1939)

INDEX